A Modern Approach To Disease Classification And Clinical Coding

For Health Information Management Students, Professionals, and Educators

Folásayò Ayégbayò, *DBA, Ph.D, FCIML*

A MODERN APPROACH

TO

DISEASE CLASSIFICATION AND CLINICAL CODING

By

FOLASAYO AYEGBAYO

DEPARTMENT OFHEALTH INFORMATION MANAGEMENT
LAGOS STATECOLLEGE OF HEALTH TECHNOLOGY, YABA
LAGOS NIGERIA

First Published in 2009
New improved Edition - 2017
Copyright ©2009, 2017: Folasayo Ayegbayo
₊234 – 0802 – 757 – 3551
ISBN 978-0-359-11949-3

DEDICATION

To the Glory of God

To the memory of my late parents, Pa Oladokun Ayegbayo and Madam Jolaade Ayegbayo

And to my beloved and caring wife, Folasade Ashabi Ayegbayo, and my lovely children.

PREFACE

The field of Health Information Management (HIM) has come through different stages of metamorphosis, with many attendant challenges. The story is also the same with the practice of Disease Classification and Clinical Coding, which is a fundamental (if not the most important) aspect of HIM. Some of the factors that have directly influenced the Disease Classification and Clinical Coding the world over are: 1. the periodic revision of the ICD, and 2. the untamable advancement in Information and Communication Technology, which has changed the face of the entire world.

This 11-chapter book is an attempt to present the current best practices in this specialty of HIM to Students, educators, and practitioners of HIM. This will help to ward off the frontier of academic and practical inadequacies present both in the training and practice areas.

Therefore, the book promises to provide the much needed companionship for both students Educators, and Practitioners, as they cannot afford not to be compliant in the present world that has become a "global village", particularly in the field of HIM. Every chapter in the book is very interesting as they examine fresh and emerging facts in the practice of Disease Classification and Clinical coding. Some of these include chapters on emerging classification; classification, nomenclature, and terminology; an introductory chapter on Billing and Reimbursement Methodologies; ethical issues in coding; future consideration and many more interesting topics.

This book does not lay any claim to perfection and any error or omission is regretted. It is simply my humble contribution towards making Disease Classification and Clinical Coding as a focal point of study more interesting and understandable to students and practitioners alike.

Dr. Folasayo Ayegbayo.

ACKNOWLEDGEMENT

God has been so good to me and He has continued to inspire and enable me in the task of teaching and writing for posterity. I am grateful for the tremendous support of my seniors, colleagues. My invaluable students in the Department of Health Information Management have been of immense help via the little but weighty words of encouragement they always offer. I am greatly indebted to them.

Two books were particularly useful in the collation of materials for this book. They are "Health Information Management" edited by LaTour, K. and Eichenwald-Maki, S; and the Medical Record book by Mogli. Special thanks also go to the World Health Organization (WHO) for making the information used in the chapter containing "ICD updates" available on its web site.

TABLE OF CONTENTS

Dedication

Preface

Acknowledgement

Table of Contents

CHAPTER ONE

FOUNDATION OF DISEASE CLASSIFICATION

INTRODUCTION

In order to ensure accurate comparison of morbidity and mortality data specified for various diseases and causes of death, it becomes necessary that a uniform method of classifying diseases and causes of death be used worldwide. Such a classification was introduced several years ago then known as the international Classification of Causes of Death. It later became known as the International Statistical Classification of Disease, Injuries and Causes of Death (ICD). Since its introduction, it has been revised once every ten years, the latest revision being the 10th Revision, which is now called International Statistical Classification of Diseases and Related Health Problems (ICD – 10).

Healthcare is faced with many challenges, including an ageing population, the need to conserve resources, medical knowledge that is increasing exponentially, and a consumer population with internet access. To meet these challenges, healthcare organizations must have the ability to operate effectively and efficiently using the latest medical data knowledge.

It is difficult to believe that the quest to classify morbidity and mortality is quite old. London parishes first began to keep death records in 1532. In 1662, John Graunt, a merchant, wrote 'Natural and Political Observation made upon the Bill of Mortality'. His friend, Sir William Petty, was able to extrapolate from mortality rates an estimate of community economic loss caused by deaths. Two hundred years later, Florence Nightingale, in 'Notes on a

Hospital', wrote "In attempting to arrive at the truth, I have applied everywhere for information, but in scarcely an instance have I been able to obtain hospital records fit for any purposes of comparison. If they could be obtained ...., they would show subscribers how their money was being spent, what amount of good was really being done with it, or whether the money was not doing mischief rather than good" (Barnett et al. 1993 p. 1046).

Many of the same issues remain in healthcare today. It is vitally important to be able to compare data for outcomes measurement, quality assurance, resource utilization, best practices, and medical research. These tasks can only be achieved only when healthcare has a common terminology that is easily integrated into the computer – based patient record.

HISTORY OF THE DEVELOPMENT OF THE ICD

EARLY HISTORY:

Sir George Knibbs, the eminent Australian Statistician, credited Francois Bossier de' Lacroix (1706 – 1777), better known as Sauvages with the first attempt to classify diseases systematically. Sauvages' comprehensive treatise was published under the 'Nosologia Methodica'. A contemporary of Sauvages was the great methodologist Linnaeus (1707 – 1778), one of whose treatise was entitled 'Genera morborum'. At the beginning of the 19th century, the classification of disease in most general use was one by William Cullen (1710 – 1790), of Edinburgh, which was published in 1785 under the title 'Synopsis nosologia methodicae'.

For all practical purposes, however, the statistical study of disease began a century earlier with the work of John Graunt on the London Bill of mortality. The kind of classification envisaged by the pioneer is exemplified by his attempt to estimate the proportion of live born children who died before reaching the age of six years, no records of age at death being available. He took all deaths classed as thrush, convulsion, rickets, teeth and worms, abortive, infants, live grown, and overlaid and added to them half the deaths classed small pox, swine pox, measles, and worms without convulsions. Despite the crudity of this classification his estimate of a 36% mortality before the age of six years appears from later evidence to have been a good one. While three centuries have contributed something to the scientific accuracy of disease classification, there are many who doubt the usefulness of attempts to compile statistics of disease, or even causes of death, because of the difficulties of classification. To these, one can quote Major Greenwood: "The scientific purist, who will wait for medical statistics until they are nosologically exact, is no wiser than Horace's rustic waiting for the river to flow away".

Fortunately for the progress of preventive medicine, the General Register office of England and Wales, at its inception in 1837, found in William Farr (1807 – 1883) its first Medical Statistician – a man who not only made the best possible use of the imperfect classification available at the time, but labored to secure better classification and international uniformity in their use.

Farr found the classification of Cullen in use in the public service of his day. It had not been revised to the advances of medical science, nor was it deemed by him to be satisfactory for statistical purposes. In the first annual report of the Registrar General, therefore, he discussed the principle that should govern a statistical classification of disease and urged the adoption of a uniform classification as follow:

The advantages of a uniform statistical nomenclature, however imperfect, are so obvious, that it is surprising no attention has been paid to its enforcement in Bills of Mortality. Each disease has, in many instances, been denoted by three or four terms, and each term has been applied to as many different diseases: vague inconvenient names have been employed, or complications have been registered instead of primary diseases. The nomenclature is of as much importance in this department of inquiry as weights and measures in the physical sciences, and should be settled without delay.

Both nomenclature and statistical classification received constant study and consideration by Farr in his annual 'letters' to the Registrar General. The utility of a uniform classification of causes of death was so strongly recognized at the first International Statistical Congress, in Brussels in 1853, that the congress requested William Farr and Marc d' Espine, of Geneval to prepare an internationally applicable, uniform classification of causes of death. At the next congress, in Paris in 1855, Farr and d'Espine submitted two separate lists which were based on very different principles. Farr's classification was arranged under five groups: epidemic diseases, constitutional (general) diseases, local diseases arranged according to anatomical sites, developmental diseases, and diseases that are the direct result of violence. D'Espine classified diseases according to their nature (gouty, herpetic, haemetic, etc).

The congress adopted a compromise list of 139 rubrics. In 1864, this classification was revised in Paris on the basis of Farr's model and was subsequently further in 1874, 1880, and 1886. Although, this classification was universally accepted, the general arrangement proposed by Farr, including the principle of classifying diseases by anatomical site, survived as the basis of the International List of Causes of Death.

ADOPTION OF THE INTERNATIONAL LIST OF CAUSES OFDEATH

The International Statistical Institute (ISI), the successor to the International Statistical Congress, at its meeting in Vienna in 1891, charged a committee chaired by Jacques Bertillon (1851 – 1922), Chief of Statistical Services of the city Paris, with the preparation of classification of causes of death. It is of interest to note that Bertillon was the grandson of Achille Guillard, a noted botanist and statistician, who had introduced the resolution requesting Farr and d'Espine to prepare a uniform classification at the first International Statistical Congress in 1853. The report of this committee was presented at the meeting of the ISI in Chicago in 1893 and adopted it. The classification prepared by Bertillon's committee was based on the classification of causes of death used by the city of Paris, which, since its revision in 1885, represented a synthesis of English, German, and Swiss Classification. The classification was based on the principle, adopted by Farr, of distinguishing between general diseases and those localized to a particular organ or anatomical site. In accordance with the instruction of the Vienna Congress made at the suggestion of L.

Guillaume, the Director of the Federal Bureau of Statistics of Switzerland, Bertillon included three classifications: the first, an abridge classification of 44 titles; the second of 99 titles and the third, a classification of 161 titles.

The Bertillon Classification of Causes of Death, as it was first called, received general approval and it was adopted by several countries, as well as by many cities. The classification was first used in North America by *Jesus E. Monjaras* for the statistics of *San Luis de Potosi, Mexico*. In 1898, the American Public Health Association, at its meeting in Ottawa, Canada, recommended the adoption of the Bertillon Classification by registrars of Canada, Mexico, and the United States of America. The Association further suggested that the classification should be revised every ten years.

At the meeting of the International Statistical Institute (ISI) at Christiania in 1899, Bertillon presented a report on the progress of the classification, including the recommendations of the American Public Health Association for *decennial revisions*. The ISI then adopted the following resolutions:

> *The International Statistical Institute convinced of the necessity of using in the different countries comparable nomenclatures:*
>
> *Learns with pleasure of the adoption by all the statistical offices of North America, by some of those of South America, and by some in Europe, of the system of causes of death nomenclature presented in 1893;*
>
> *Insists vigorously that this system of nomenclature be adopted in principle and without revision, by all the Statistical Institutions of Europe;*

Approves, at least in its general lines, the system of decennial revision proposed by the American Public Health Association at its Ottawa session (1898);

Urges the statistical offices who have not adhered, to do so without delay, and to contribute to the comparability of the causes of death nomenclature.

The French Government therefore convoked in Paris, in August 1900. The desirability of decennial revision was recognized, and the French Government was requested to call the next meeting in 1910. In fact the next conference was held in 1909, and the Government of France called succeeding conferences in 1920, 1929, and 1938.

Bertillon continued to be the guiding force in the promotion of the International List or Causes of Death, and the revisions of 1900, 1910, and 1920 were carried out under his leadership. As Secretary – General of the International Conference, he sent out the provisional revision for 1920 to more than 500 people, asking for comment. His death in 1922 left the International Conference without a guiding hand.

At the 1923 session of the ISI, Michel Huber, Bertillon's successor in France, recognized this lack of leadership and introduced a resolution for the ISI to renew its stand of 1893 in regard to the International Classification of Causes of Death and to cooperate with other International Organizations in preparation for the subsequent revisions. The Health Organizations of the League of Nations had also taken an active interest in vital statistics and appointed a commission of statistical Experts to study the classification of diseases and causes of death, as well as other problems in the field of medical statistics. E. Roesle, Chief of Medical Statistical Service of the German Health Bureau and a member of the Commission of Experts statisticians, prepared a monograph that listed the expansion in the rubrics of the 1920 International List of Causes of Death. That would be required if the classification was to be used in the tabulation of statistics of morbidity. This careful study was published by the Health Organization of the League of Nations in 1928. In order to coordinate the work of both agencies, an international commission, known as the 'Mixed Commission' was created with an equal number of representatives from the ISI and the Health Organization of the League of Nations. This Commission drafted the proposal for the fourth (1929) and the fifth (1938) revisions of the International List of Causes of Death.

THE SIXTH REVISION OF THE INTERNATIONAL LISTS

The International Health Conference held in New York City in June and July 1946 entrusted the Interim Commission of the World Health Organization (WHO) with the responsibility of:

> *Reviewing the existing machinery and of undertaking such preparatory work as may be necessary in connection with:*
>
> 1. *The next decennial revision of 'The International List of Causes of Death' (including the lists adopted under the International Agreement of 1934, relating to Statistics of Causes of Death); and*
> 2. *The establishment of International List of Cause of Morbidity*

To meet this responsibility, the Interim Commission appointed the Expert Committee for the preparation of the Sixth Decennial Revision of the International Lists of Diseases and Causes of Death.

The Committee taking full account of prevailing opinion concerning morbidity and mortality classification review and revised the above mentioned proposed classification which had been prepared by the United States Committee on Joint Causes of Death, chaired by Lowell J. Reed, Professor of Biostatistics at Johns Hopkins University.

The resulting classification was circulated to national government preparing morbidity and mortality statistics for comments and suggestions under the title, **International Statistical Classification of Diseases, Injuries and Causes of Death (ICD).** The Expert Committee considered the replies and prepared a revised version incorporating such changes as appeared to improve the utility and acceptability of the classification. The Committee also compiled a list of diagnostic terms to appear under each title of the classification. Furthermore, a sub-committee was appointed to prepare a comprehensive alphabetical index of diagnostic statements classified to the appropriate category of the classification.

The Committee also considered the structure and was of special lists of causes for tabulation and publication of morbidity and mortality statistics and studied other problems related to the international comparability of mortality statistics, such as form of medical certificate and rules for classification.

The International Conference for the Sixth Revision of the International List of Diseases and Causes of Death was convened in Paris from 26 – 30 April, 1948 by the Government of France under the terms of the agreement signed at the close of the Fifth Revision Conference in 1938. Its secretariat was entrusted jointly to the competent French authorities and to the WHO, which had carried out the preparatory work under the terms of the arrangement concluded by the governments represented at the International Health Conference in 1946.

In 1948, the first World Health Assembly endorsed the report of the Sixth Revision Conference and adopted WHO Regulations No. 1, prepared on the basis of the recommendations of the conference. The International Classification, including the Tabular List of inclusion defining the content of the categories was incorporated, together with the form of the medical certificate of causes of death, the rules for classification and the special lists for tabulation, into the ***Manual of the International Statistical Classification of Diseases, Injuries, and Causes of Death.*** The manual consisted of two volumes, volume 2 being an alphabetical index of diagnostic terms coded in the appropriate categories. The Sixth Decennial Revision marked the beginning of a new era in international vital and health statistics. Apart from approving a comprehensive list for both mortality and morbidity and agreeing on international rules for selecting the underlying causes of death, it recommended the adoption of a comprehensive programme of international cooperation in the field of vital and health statistics. An important item in this programme was the recommendation that governments establish national committees on vital and health statistics to coordinate the statistical activities in the country, and to serve as a link between the national statistical institutions and the WHO. It was further envisaged that such national committees would, either singly or in cooperation with other national committees, study statistical problems of public health importance or make the results of their investigations available to WHO.

THE SEVENTH AND EIGHTH REVISIONS

The International Conference for the Seventh Revision of the ICD was held in Paris under the auspices of WHO in February, 1955. In accordance with a recommendation of the WHO Expert Committee on Health Statistics, this revision was limited to essential changes and amendments of errors and inconsistencies.

The Eight Revision Conference convened by WHO met in Geneva, from 6 to 12 July 1965. This revision was more radical than the Seventh but left unchanged the basic structure of the classification and the general philosophy of classifying diseases, whenever possible, according to their etiology rather than a particular manifestation.

During the years that the Seventh and Eighth Revisions of the ICD were in force, the use of ICD for indexing hospital medical records increased rapidly and some countries prepared national adaptations which provided additional detail needed for this application of the ICD.

THE NINTH REVISION

The International Conference for the Ninth Revision of the ICD, convened by WHO, met in Geneva from 30th September to 6th October 1975. In the discussions leading up to the conference, it had originally been intended that there should be a little change other than updating of the classification. This was mainly because of the expense of adapting data-processing systems each time the classification was revised. There had been an enormous growth of interest in the ICD and ways had to be found of responding to this, partly by modifying the classification itself and partly by introducing special coding provisions. A number of representations were made by specialist bodies which had become interested using the ICD for their own statistics. Some subject areas in the classification were regarded as inappropriately arranged and there was considerable pressure for more detail and for adaptations of the classification to make it more relevant for the evaluation of medical care, by classifying conditions to the chapters concerned with the part of the body affected rather than to those dealing the underlying generalized disease. At the other end of the scale, there were representations from countries and areas where a detailed and sophisticated classification was irrelevant, but which nevertheless needed a classification based on ICD in order to assess their progress in healthcare and in the control of diseases.

The final proposals presented to and accepted by the conference retained the basic structure of the ICD, although, with much additional detail at the level of the four-digit subcategories, and some optional five-digit subdivisions. For the benefit of users not requiring such detail, care was taken to ensure that the categories at the three-digit level were appropriate. For the benefit of users wishing to produce statistics and indexes oriented toward medical care, the Ninth Revision include an optional alternative method of classifying diagnostic statements, including information about both an underlying general disease and a manifestation in a particular organ or site. This system became known as the dagger and asterisk system and is retained in the Tenth Revision. A number of other innovations were included in the Ninth Revision, aimed at increasing its flexibility for use in a variety of situations.

The Twenty-ninth World Health Assembly, noting the recommendations of International Conference for the Ninth Revision of the ICD, approved the publication, for trial purposes, of supplementary classification of Impairments and Handicaps and of Procedures in Medicine as supplements, but not as integrated parts of, the ICD. The conference also made recommendations on a number of related technical subjects: coding rules for mortality were amended slightly and rules for the selection of a single cause for tabulation of morbidity were introduced for the first time; definitions and recommendations for statistics in the field of perinatal mortality were amended and extended and a certificate of causes of perinatal mortality was recommended; countries were encouraged to do further work on multiple condition coding and analysis, but no formal methods were recommended; and a new basic tabulation list was produced.

PREPARATIONS FOR THE TENTH REVISION

Even before the conference for the Ninth Revision, WHO had been preparing for the Tenth Revision. It had been realized that the great expansion in the use of the ICD necessitated a thorough rethinking of its structure and an effort to devise a stable and flexible classification, which should not require fundamental revision for many years to come. The WHO collaborating centers for Classification of Diseases were consequently called upon to experiment with models of alternative structures for ICD – 10.

It had also become clear that the established ten-year interval between revisions was too short. Work on the revision process had to start before the current version of the ICD had been in use long enough to be thoroughly evaluated, mainly because the necessity to consult so many countries and organizations made the process a very lengthy one. The Director-General of WHO therefore wrote to the Tenth Revision Conference, which was originally scheduled for 1985 and to delay the introduction of the Tenth Revision which would have been due in 1989. In addition to permitting experimentation with alternative models for the structure of the ICD, this allowed time for the evaluation of ICD – 9, for example through meetings organized by some of the WHO Regional Offices and through a survey organized at headquarters.

CHAPTER TWO

FUNDAMENTALS OF CLASSIFICATION AND NOMENCLATURE

PURPOSE AND APPLICATIONS:

William Farr in 1856, described classification thus:

"Classification is a method of generalization. Several classifications may therefore be used with advantage; and the physician, the pathologist, or the jurist, each from his own point of view, may legitimately classify the diseases and causes of death in the way he thinks best adapted to facilitate his inquiries and to yield general result".

A classification of disease has also been defined as a system of categories to which morbid entities are according to established criteria. There are many possible choices for these criteria. There are many possible choices for these criteria. The anatomist, for example may desire a classification based on parts of the body affected. The pathologist may be interested in the nature of disease process. The public health practitioner could be interested in the cause of diseases, while the clinician may be interested in the in the particular manifestation requiring his care.

Consequently, there are many axes of classification. They include:

a. **Anatomical or Topographical axis** – this is the classification or arrangement of diseases according to the part or organ of the body affected.
b. **Etiological axis** – this is the classification of diseases according to causes and/or causative factors

c. **Morphological axis** – this is the classification of diseases according to morbid behavior produced by the disease or injury.

d. **Functional axis** – this is the classification of diseases according to the kind of disturbance of function produced by the disease or injury.

e. **Alphabetical axis** – this is the classification of diseases according to the arrangement of alphabetical order.

f. **Epidemiological axis** – this is the classification of diseases in accordance with the trends of the disease within a specific population, with the primary aim of finding the causes and effects of diseases and conditions. This takes factors such as incidence (new cases) and prevalence (old and new cases) of such disease into consideration.

g. **Pathological axis** – the axis of classification concerned with the changes caused in the body by the disease process.

h. **Clinical axis** – the axis of classification concerned with the way the disease manifests itself.

However, the purpose of the ICD is to permit the systematic recording, analysis, interpretation, and comparison of mortality and morbidity data collected in different countries or areas and at different times. The ICD is used to translate diagnoses of disease and other health problems from words into an alphanumeric code, which permit easy storage, retrieval, and analysis of the data.

In practice, the ICD has become the international standard diagnostic classification for all general epidemiological and many health management purposes. These include the analysis of the general health situation of population groups and the monitoring of the incidence and prevalence of diseases and other health problems in relation to other variables such as the characteristics and circumstances of the individuals affected. The ICD is neither intended nor suitable for indexing of distinct clinical entities. There are also some constraints on the use of the ICD for studies of financial aspect, such as billing or resource allocation. These constraints can be overcome as in the case of the ICD – 10 – CM in use in the United States of America.

The ICD can be used to classify diseases and other health problems recorded on many types of health and vital records. Its original use was to classify causes of mortality as recorded at the registration of death. Later its scope was extended to include diagnoses in morbidity. It is important to note that, although the ICD is primarily designed for the classification of diseases and injuries with a formal diagnosis, not every problem or reason for coming into contact with health services can be categorized in this way. Consequently, the ICD provided for a wide variety of signs, symptoms, abnormal findings, complaints, and social circumstances that may stand in the place of a diagnosis on health-related records. It can therefore be used to classify data recorded under headings such as 'diagnosis' 'reason for admission' 'conditions treated' and 'reason for consultation', which appear on a wide variety of health records from which statistics and other health-situation information are derived.

INTERNATIONAL NOMENCLATURE OF DISEASES (IND)

In 1975, the IND became a joint project of the Council for International Organizations of Medical Sciences (CIOMS) and World Health Organization (WHO), guided by a Technical Steering Committee of representatives of both organizations.

The principal objective of IND is to provide, for each morbid entity a single recommended name. The main criteria for selection of this name are that it should be specific (applicable to only one disease), unambiguous, as self- descriptive as possible, as simple as possible, and (whenever feasible) based on cause. However, many widely used names that do not fully meet the above criteria are being retained as synonyms, provided they are not inappropriate, misleading, or contrary to the recommendations of International Specialist Organizations. Eponymous terms are avoided since they are not self-descriptive; however, many of these names are in such widespread use (e.g. Hodgkin disease, Parkinson disease, and Addison disease) that they must be retained.

Each disease or syndrome for which a name is recommended is defined as unambiguously and as briefly as possible. A list of synonyms appears after each definition. These comprehensive lists are supplemented, if necessary by explanations about why certain synonyms have been rejected or why an alleged synonym is not a true synonym.

The IND is intended to be complementary to the ICD. As far as possible, IND terminology has been given preference in the ICD. The volumes of the IND published up to 1992 are:

1. *Infectious diseases {bacteria diseases [1985]; mycoses [1982]; viral diseases [1983]; parasitic diseases [1987]}*
2. *Diseases of the lower respiratory tract [1979]*
3. *Diseases of the digestive system [1990]*
4. *Cardiac and vascular diseases [1989]*
5. *Metabolic, nutrition, and endocrine disorders [1991]*
6. *Diseases of the kidney, the lower urinary tract, and the male genital system [1992]*
7. *Diseases of the female genital system [1992]*

DIFFERENCES BETWEEN A NOMENCLATURE AND A CLASSIFICATION

A statistical classification of diseases must be confined to a limited number of mutually exclusive categories able to encompass the whole range of morbid conditions. The categories have to be chosen to facilitate the statistical study of disease phenomena. A specific disease entity that is of particular public health importance or that occurs frequently should have its own category. Otherwise, categories will be assigned to groups of separate but related conditions. Every disease or morbid condition must have a well defined place in the list of categories. Consequently, throughout the classification, there will be residual categories for others and miscellaneous conditions that cannot be allocated to the more specific categories. As few conditions as possible should be classified to residual categories.

It is the element of grouping that distinguishes a statistical classification from a nomenclature, which must have a separate title for each known morbid condition. The concepts of classification and nomenclature are nevertheless closely related because a nomenclature is often arranged systematically.

A statistical classification can allow for different levels of detail if it has hierarchical structure with subdivisions. A statistical classification of diseases should retain the ability both to identify specific disease entities and to allow statistical presentation of data for broader groups to enable useful and understandable information to be obtained.

The same general principle can be applied to the classification of other health problems and reason for contact with healthcare services, which are also incorporated in the ICD.

The ICD has developed as a practical, rather than a purely theoretical classification, in which there are a number of compromises between classifications based on etiology, anatomical site, circumstances of onset, etc. there have also been adjustments to meet the variety of statistical applications for which the ICD is designed, such as mortality, morbidity, social security and other types of health statistics and surveys.

TYPES OF NOMENCLATURE

Generally speaking, *Nomenclature is described as a recognized system of terms used in a science or art that follows pre-established naming convention.* However, in the field of medicine, nomenclature can be of different types, these are:

1. **Disease Nomenclature** – a listing of the proper names for each disease entity with its specific code number.

2. **Procedure or Operation Nomenclature** – a listing of proper names for operative procedures with their specific code numbers.

3. **Drug or Prescription Nomenclature** – a listing of proper names of drugs, chemicals, and prescription terms with appropriate code numbers, as endorsed by professional and regulatory agencies.

4. **Equipment Nomenclature** – a listing of proper names or terms describing various medical and surgical equipment as endorsed by manufacturer, professional, and standards regulatory bodies.

5. **Laboratory Nomenclature** – a listing of approved proper names for laboratory tests, and results with their respective numbers, e.g. LOINC.

6. **Nursing Practice Nomenclature** – an approved list of common terms describing nursing practice, containing nursing phenomena, nursing actions, and nursing outcomes.

CHAPTER THREE

CLASSIFICATION SYSTEMS FOR HEALTHCARE DATA

INTRODUCTION:

As the discussion of classification systems for healthcare data begins, it is important to have a common understanding of how **Classifications, Nomenclatures**, and **Terminologies** are defined. Unfortunately, it is difficult to get complete agreement in even this area. The most widely used accepted terminology standard is the International Standards Organization (ISO) Standard 1087 (Terminology-Vocabulary) (Hammond and Cimino 2000, p. 224)

When health care providers document patient care they use a medical nomenclature which is a vocabulary of clinical and medical terms (e.g., myocardial infarction, diabetes mellitus, appendectomy, and so on). A coding system (or classification system) organizes a medical nomenclature according to similar conditions, diseases, procedures, and services and establishes codes (numeric and alphanumeric characters) for each. Codes are reported to third-party payers for reimbursement, to external agencies for data collection, and internally for education and research.

EXAMPLE

ICD-10 codes are assigned to hospital inpatient diagnoses and procedures and entered into automated abstracting software. Codes are reported to the billing department and printed on inpatient bills. Diagnosis and procedure indexes are generated from abstracted data and later used to retrieve specific patient records and summary data for cancer research.

AHIMA Coding Policy and Strategy Committee also defines Clinical Vocabulary as a list or collection of clinical words or phrases with their meanings. In the committee's opinion, a Clinical Terminology provides for the proper use of clinical words as names and symbols. The committee equated a clinical terminology with a nomenclature (AHIMA 1999, p. 72). It is important to recognize that this problem of multiple definitions and names is endemic in the field of healthcare terminology. Standardized terms can be efficiently mapped to broader classification for administrative, regulatory, oversight, and fiscal requirements (Chute 2000, p. 301).

WHAT IS CLASSIFICATION?

It is crystal clear that there are several ways by which one can define classification, particularly when it is being considered from different perspectives and contexts. In view of the present day realities, the present context of disease classification, and the various classification systems available globally for different applications, a classification can therefore be defined as:

'A clinical vocabulary, terminology, or nomenclature that lists words or phrases with their meanings, provides for the proper use of clinical words as names or symbols, and facilitates mapping standardized terms to broader classifications for administrative, regulatory, oversight and fiscal requirement'.

A Classification System can be defined thus:

1. A system for grouping similar diseases and procedures and organizing related information for easy retrieval.

2. A system for assigning numeric or alphanumeric code numbers to represent specific diseases and/or procedures.

Nomenclatures (in order of development)

	Nomenclature	Description

1	*Basle Nomina Anatomica*
	
	Standardized Nomenclature of Diseases (SND)
		• Developed in the late 1800s by the Anatomical Society
		
		• Initiated standardization of anatomical terms used in medicine
		
		• Developed in 1929 by the New York Academy of Medicine
		
		• First medical nomenclature to be universally accepted in the United States
		
		• Introduced the concept of multi-axial coding:
		
		  • Topology (anatomy)
		
		  • Etiology (cause of disease)
		
		NOTE: Although entitled a nomenclature, SND is also a classification system.
		
		EXAMPLE: Prostate is assigned code 764. Adenocarcinoma is assigned code 8091

		Adenocarcinoma of the prostate is coded as 764-8091.
2	Standardized Nomenclature of Diseases and Operation (SNDO)	• Developed in 1936 by the American Medical Association (AMA) • Based on SND, and added an axis for operations **EXAMPLE:** Radical excision is assigned code 14. Prostate is assigned

		code 764. Radical prostatectomy is coded as 764-14.
3	Systematized Nomenclature of Pathology (SNOP)	• Published in 1965 by the College of American Pathologists (CAP) • Four-axis system of terms and related codes for use by pathologists interested in storage and retrieval of medical data
4	Systematized Nomenclature of Medicine (SNOMED)	• Developed in 1974 by CAP, based on SNOP, and cross-referenced to ICD-9-CM • Codifies all activities within the patient record, including medical diagnoses and procedures, nursing diagnoses and procedures, patient signs and symptoms, occupational history, and the many causes and etiologies of diseases (e.g., infectious conditions, genetic and congenital conditions, physical causes of injury • CAP publishes subsequent revisions: • SNOMED II (1979) (hierarchy expanded to 6 modules)

		• SNOMED III (1993) (hierarchy expanded to 11 modules)
		• SNOMED DICOM Microglossary (SDM) (1996) (collaborative effort between CAP and the American College of Radiology and the National Equipment Manufacturers Association; developed as a subset of the Digital Imaging and Communications in Medicine [DICOM] standard)
		•• SNOMED® RT (2000) (collaborative Reference Terminology effort with the Kaiser Permanente Convergent Medical Technology [CMT] Project; integrated with Logical Observation Identifiers Names and Codes [LOINC®] a universal standard medical vocabulary for identifying laboratory and clinical observations)
		NOTE: The laboratory section of LOINC covers chemistry, hematology, microbiology, serology, and toxicology. The clinical observation section contains entries for cardiac echocardiography, EKG, gastroendoscopic procedures, hemodynamics, intake/output,

obstetric ultrasound, pulmonary ventilator management, urologic imaging, and vital signs.

•• SNOMED CT® (2002) (combines content and structure of SNOMED RT® with U.K. National Health Service's Clinical Terms Version 3) (formerly Read Codes, which were developed in the early 1980s by Dr. James Read to record and retrieve primary care data in a computer)

NOTE: The National Library of Medicine (NLM) will provide free-of-charge access through its Unified Medical Language System® (UMLS®) Metathesaurus® to SNOMED CT® core content and all version updates, starting in January 2004.Qualifying entities include U.S. federal agencies, state and local government agencies, territories, the District of Columbia, and any public, for-profit, and nonprofit organization located, incorporated, and operating in the United States.

5	Current Medical Information & Terminology (CMIT)	•Published in 1981 by the AMA • Used for naming and describing diseases and conditions in practice and in areas related to medicine **NOTE:** No subsequent revisions were published.
6	Unified Medical Language System® (UMLS®)	• National Library of Medicine (NLM) research and development project begun in 1986 (and remains ongoing as a long-term project) • Purpose is to aid in development of systems to help health professionals and researchers retrieve and integrate electronic biomedical information from a variety of sources and to make it easy for users to link disparate information systems, including computer-based patient records, bibliographic databases, factual databases, and expert systems

EVOLUTION OF NOMENCLATURES CLASSIFICATIONS

Systems for classifying diseases have progressed through various stages since the first classification system (Bertillon's Classification) was developed in the late 19th Century. The following sections describe current and near- future developments in the classifications.

Medical Classification and Coding Systems (in order of development)

	Coding System	Description
1	London Bills of Mortality	• Developed during the latter part of the sixteenth century • Considered the first classification system • Bills were collected and collated by parish clerks (with no medical training)
2	*Nosologia Methodica*	• Medical classification system developed in the mid-1700s by François Bossier de Lacroix (Sauvages)

3	Bertillon International Statistical Classification of Causes of Death	• Classification of diseases by site, adopted in 1893
		• Subsequent revisions were entitled *International List of Causes of Death* (ICD-1, ICD-2, ICD-3, and ICD-4)
		• Classifications of diseases for morbidity reporting purposes were integrated into subsequent revisions
		• ICD-5 added mental diseases and deficiency (mental deficiency, schizophrenia, manic depressive psychosis, and other mental diseases)
		• In 1946, the World Health Organization (WHO) revised ICD-6 and established an International List of Causes of Morbidity

| 4 | Manual of the International Statistical Classification of Diseases, Injuries and Causes of Death (ICD-6) | • ICD-6 is adopted internationally in 1948 by the First World Health Assembly

,• WHO reviews and revises ICD about every 10 years:

• • ICD-7 (1955)

• • ICD-8 (1965)

• • ICD-9 (1975)

• • ICD-10 (1989, with WHO member states adopting in 1994)

• ICD-10 is entitled the *International Statistical Classification of Diseases and Related Health Problems (ICD-10)* and differs from ICD-9:

• • More detailed (8,000 categories vs. 4,000 in ICD-9)

• • Uses three-digit alphanumeric category codes (vs. three-digit numeric category codes in ICD-9)

• • Contains three additional chapters, and other chapters have been reorganized |

		• • Cause-of-death titles are modified and conditions reorganized
		• • Some coding rules have changed
		• • Published in three volumes (vs. two volumes in ICD-9)
		NOTE: The United States has commenced the process of full implementation of ICD-10-CM and ICD-10-PCS in 2015.
5	International Classification of Diseases, Adapted for Use in the US (ICDA)	• The United States adapted ICD-8 in 1966 to include additional detail for coding hospital and morbidity data, abbreviated ICDA-8.
		• In 1968, the Commission on Professional and Hospital Activities (CPHA) of Ann Arbor, Michigan, published a hospital adaptation of ICDA, entitled H-ICDA, which was revised in 1973 as H-ICDA-2.
		NOTE: United States hospitals were divided in their use of ICDA-8 and H-ICDA. (Author Green recalls coding inpatient cases using ICD-9-CM in 1979, while her utilization review coordinator coded according to H-ICDA as required by the

| | | county Professional Standards Review Organization.)

• In 1979, all hospitals were required to adopt the International Classification of Diseases, Ninth Revision, Clinical Modification (ICD-9-CM), which classifies

diagnoses (Volumes 1 and 2) and procedures (Volume 3). All hospitals and ambulatory care settings use ICD-9-CM to report diagnoses; hospitals use ICD-9-CM procedure codes to report inpatient procedures and services.

• In 2013, the United States will implement ICD-10-CM and ICD-10-PCS. The National Center for Health Statistics (NCHS) is the federal agency responsible for developing ICD-10-CM. The ICD-10-PCS (Procedure Coding System) was developed with support of the Centers for Medicare & Medicaid Services under contract with 3M Health Information Systems. The National Committee on Vital Health and Statistics (NCVHS) serves as a public |

		advisory body to the Secretary of DHHS in the development of ICD-10-PCS.
6	Diagnostic and Statistical Manual Mental Disorders (DSM)	• Published by the American Psychiatric Association as a standard classification of mental disorders used by mental health professionals in the UnitedStates
		• • DSM (1952)
		• • DSM-II (1968)
		• • DSM-III (1980) and a multi-axial classification was added:
		• • • Axis I—mental disorders or illnesses (e.g., substance abuse)
		• • • Axis II—personality disorders or traits (e.g., mental retardation)
		• • • Axis III—general medical illnesses (e.g., hypertension)
		• • • Axis IV—life events or problems (e.g., divorce)
		• • • Axis V—global assessment of functioning (GAF) (e.g., occupational)
		• • DSM-III-R (1987)

		• • DSM-IV (1994) • • DSM-IV-TR (2000) (Text Revision to correct DSM-IV errors, update codes according to ICD-9-CM annual revisions, and so on) • • DSM-V (expected in 2013) • Derived from ICD-9-CM, designed for use in a variety of health care settings, and consists of three major components: • • Diagnostic classification • • Diagnostic criteria sets • • Descriptive text
7	Current Procedural Terminology (CPT)	• Originally published by the American Medical Association in 1966. • Subsequent editions were published about every five years, until the late1980s when the AMA began publishing annual revisions of CPT-4. • The CPT-5 Project was initiated by the AMA in 2000 to address challenges presented by emerging user needs, the Health Insurance

Portability and Accountability Act of 1996 (HIPAA), and needed improvements in CPT. The

primary goal of the CPT-5 Project was to have CPT chosen by the Secretary of Health and Human Services as the national standard procedure code set for physician services under HIPAA. A Final Rule, issued in the August 17, 2000, *Federal Register,* named CPT as the national standard code set for physician services. CPT-5 Project recommendations are implemented annually along

with code additions/ deletions/revisions.

• CPT classifies procedures and services; physicians and ambulatory care settings use CPT codes to report procedures and services.

• CPT is level I of the Healthcare Common Procedure Coding System (HCPCS).

8	International Classification of Diseases for Oncology, third edition (ICD-O-3)	• First edition of ICD-O was published in 1976, and a revision (primarily of topography codes) was published in 1990.
		• ICD-O-3 was implemented in 2001.
		• Ten-digit code, which describes the tumor's primary site (four-characters) topography code, histology (four-digit cell type code), behavior (one-digit code for malignant, benign, and so on), and aggression (one-digit differentiation or grade code).
9	International Classification of Injuries, Disabilities, and Handicaps (ICIDH)	• Published in 1980, ICIDH classifies health and health-related domains that describe body functions and structures, activities, and participation.
		• ICF complements ICD-10, looking beyond mortality and disease.
		• In 2001, with publication of its second edition, the name changed to *International Classification of Functioning, Disability and Health* (ICF).

10	HCPCS Level II (national codes)	• Published by a variety of vendors, the coding system is in the public domain, which means it is not copyrighted. • Managed by the Centers for Medicare & Medicaid Services (CMS). • Classifies medical equipment, injectable drugs, transportation services, and other services not classified in CPT. Physicians and ambulatory care settings use HCPCS Level II to report procedures and services.
11	Current Dental Terminology (CDT)	• Published biannually by the American Dental Association (ADA). • Classifies dental procedures and services. • Dental providers and ambulatory care settings use CDT to report procedures and services. • CDT codes are also included in HCPCS Level II; they begin with the alphacharacter D.

1 3	National Drug Codes (NDC)	• Published by a variety of vendors, the coding system is in the public domain. • Managed by the Food and Drug Administration (FDA). • Originally established as part of an out-of-hospital drug reimbursement program under Medicare. • Serves as a universal product identifier for human drugs. • Current edition is limited to prescription drugs and a few selected overthe-counter (OTC) products. • Retail pharmacies use NDC to report pharmacy transactions. • Some health care professionals also report NDC codes on claims.
1 3	Alternative Billing Codes (ABCcodes™) •	• Alternative Billing Codes (ABCcodes™) classify services not included in theCPT manual to describe the service, supply, or therapy provided; they may also be assigned to report nursing services and alternative medicine

professions. The codes are five characters in length, consisting of letters, and they are supplemented by two-digit code modifiers to identify the practitioner

performing the service.

EXAMPLE: During an office visit, an acupuncture physician assessed the health status of a new client and developed a treatment plan, a process that took 45 minutes. ABCcode assigned: ACAAC-1C. (The office visit is coded as ACAAC, and the acupuncture physician is assigned modifier 1C.)

• HIPAA authorized the Secretary of DHHS to permit exceptions from HIPAA transaction and code set standards to commercialize and evaluate proposed

modifications to those standards. The ABCcodes™ system was granted that exception in 2003, and the codes were evaluated through 2005 for possible adoption. However, adoption was not approved.

ADAPTATION OF THE ICD-10

Applications completed or in development include:

- Dentistry and Stomatology ICD-DA
- Dermatology
- Mental and behavioral disorders (already available in several languages)
- Neurology (ICD-10 NA)
- Oncology (ICD-O-2, is already available in several languages and implemented in a number of countries)
- Pediatrics
- Psychiatry in primary care
- Rheumatology and Orthopedics (ICD-R&O)

Plans are well advanced for further specialty-based applications of ICD-10, these being:

- Hereditary diseases with the unit of Human Genetics
- External causes with the WHO Safety Promotion and Injury Control Unit and Nordic Medico-statistical Committee
- International Classification of Impairments, Disabilities and Handicaps (ICDDGýH)

Revisions are taking place in the following areas:

- Handicaps – revising codes for the handicap section, as well as how to define "handicap" in a way that distinguishes it from "disability"
- Environmental factors – insuring that the social and environmental barriers affecting disability and handicap are appropriately recognized in proposed new classifications

- Child disabilities/mental and behavioural health/cognitive and behavioural disabilities disability policy – task forces in these areas are changed with insuring the new classifications of I, D and H are useful for applications in programmes for children and persons with mental illness, and in social policy.

CHAPTER FOUR

EMERGING VOCABULARIES AND CLASSIFICATIONS

INTRODUCTION

A number of vocabularies and classifications have been developed in recent years, some of which will have a significant impact on the role of HIM professionals in the future. This section examines some of these recent and emerging vocabularies and classifications.

SYSTEMATIZED NOMENCLATURE OF MEDICINE

The Systematized Nomenclature of Medicine (SNOMED) is developed, maintained, and distributed by the College of American Pathologist (CAP). SNOMED was originally built on the Systematized Nomenclature of Pathology (SNOP), which was introduced in 1965. Like SNOP, SNOMED was an alphanumeric, multi-axial coding scheme. (Kudla and Blakemore 2001).

In 1997, the CAP worked with a team of physicians and nurses from ***Kaiser Permanente*** to begin development of the Systematized Nomenclature of Medicine Reference Terminology (SNOMED-RT). (Kudla and Blakemore 2001). In 1997, Spackman, Campbell, and Cote defined a reference terminology for clinical data as "a set of concepts and relationships that provides a common reference point for comparison and aggregation of data about the entire healthcare process, recorded by multiple different individuals, systems or institutions". One of the ways that SNOMED-RT did this was by including an elementary mapping to ICD-9-CM. (Kudla and Blakemore 2001). SNOMED also worked with the Digital Imaging and Communications in Medicine (DICOM) Community, the Logical Observation Identifier Names and Codes

(LOINC) System, and the various nursing vocabularies to further expand its content.

In 2002, SNOMED-RT and Clinical Terms, Version 3 (CTV3), also known as the Read Codes, merged to create the SNOMED-Clinical Terminology (CT) system. (NHS 1999). This merge more than double the content of the system to more than 300,000 unique clinical concepts and more than 900,000 descriptions to express these concepts. SNOMED-CT provides coverage of the following:

- Disease
- Procedure
- Body structure
- Organism
- Finding
- Observable entity
- Specimen
- Substance
- Pharmaceutical/biologic product
- Physical object
- Physical force
- Measurable
- Events
- Social context
- Environmental and geographical locations
- Staging and scales.

SNOMED-CT also provides cross mapping to:

- ICD-10 topography
- ICD-9-CM

- ICD-10
- LOINC
- Nursing Classifications PNDs, NANDA, and Omaha (Ryske and Imel 2003)

LOGICAL OBSERVATION IDENTIFIER NAMES AND CODES (LOINC)

The work on LOINC began in February, 1994. Today LOINC is generally accepted as the exchange standard for laboratory result, enabling standards to be developed and adopted relatively quickly to meet a desperate need. The *Regenstrief Institute* maintains the LOINC database and its supporting documentation.

The goal of LOINC is not to replace the Laboratory field in facility databases, but rather to provide mapping mechanism. The LOINC committee hoped that laboratories would create fields in the master files for storing LOINC codes and names as attributes of their own data elements.

Each LOINC name is structured and can contain up to six parts, including:

- Analyte/component (for example, potassium, hemoglobin, etc)
- Kind of property measured or observed (for example, mass, mass concentration, enzyme concentration)
- Time aspect of the measurement or observation (a point in time versus an observation integrated over time)
- System/sample type (for example, urine, blood, serum)
- Type of measurement or observation scale {quantitative [a number] versus qualitative [a trait such as cloudy]}

- Type of measurement or observation method used (for example, clean catch or catheter)

The primary disadvantage of LOINC is that it may require significant modifications to work with a current laboratory information system. As with SNOMED-RT and CTV3, LOINC is usable only in computerized system.

A distinct advantage to using LOINC is that it enables the standardized communication of laboratory results. Large integrated delivery systems that have very diverse laboratory processing systems (the machines that perform the task) will find it easier to maintain and use a CPR with valid laboratory results.

NURSING CLASSIFICATIONS

The care delivered by nurses is very different from the care delivered by other healthcare professionals. In an effort to research, teach, practice, and create public policy for nursing, the nursing profession under the leadership of the American Nursing Association (ANA), has developed nursing classifications. The purpose of these classifications is to describe the nursing process, document nursing care, and facilitate aggregation of data for comparison at the local, regional, national, and international levels (Henry *et al.* 1998)

The ANA established the Steering Committee on Databases (SCD) to support Clinical Nursing Practice to monitor and support the development and evolution of the use of multiple vocabularies and classification schemes (Henry *et al.*1998). The Nursing Information and Data Set Evaluation Center (NISDEC) arose from that initiative. The purpose of the NISDEC is to review, evaluate against defined criteria, and recognize IS developers and manufacturers that support nursing care documentation with automated Nursing Information System (NIS) or within CPR system. Through the NISDEC, the ANA has recognized a total of classifications for nursing.

Currently, nursing professionals are working with SNOMED-RT and other vocabulary system developers to create classifications that will meet the needs of nursing. As Henry states in the May 1998 Journal of the American Health Information Management Association:

"In this era of numerous requests for data and information from multiple accrediting, governing, and quality monitoring agencies, it is vital that HIM professional be aware of classification systems and related national efforts, beyond those that are typically physician-centric in nature (e.g. ICD-9-CM, and CPT). Without reliable and valid data concerning the contributions of the entire healthcare team, it is truly impossible to engage in the practice of evidence-based healthcare delivery"

❖ NORTH AMERICAN NURSING DIAGNOSIS ASSOCIATION

North American Nursing Diagnosis Association (NANDA II) is a nursing terminology used to develop and classify nursing diagnoses in taxonomy. It was recognized by the ANA in 1991, added to NLM's Metathesaurus in 1993, registered by HL7 in 2000 and licensed by SNOMED in 2002 (NAHIT 2005)

❖ NURSING INTERVENTION CLASSIFICATION

Nursing Intervention Classification (NIC) is a terminology that describes the treatments that nurses perform. It contains a standardized list of 514 interventions. The classification is maintained by the Center for Nursing Classification and Clinical Effectiveness. The University of Iowa College of Nursing (Dochterman 2004)

Each intervention has a definition, a set of activities that the nurse performs to carry out the intervention, and a short list of background readings. Interventions are organized in a three level taxonomic structure that makes it easier to select an intervention and use on a computer. It is useful for clinical documentation, communication of care across settings, integration of data across systems, and curricular design. NIC was recognized by the ANA in 1990, included in the JCAHO's chapter management of information in 1994, was registered in 2000 by HL7 and licensed by SNOMED in 2002 (NAHIT 2005)

❖ NURSING OUTCOMES CLASSIFICATION

Nursing Outcomes Classification (NOC) is a comprehensive standardized classification of patient/client outcomes developed to evaluate the effects of nursing interventions. It contains a list of 330 standardized outcomes, each with a definition, a list of indicators that can be used to evaluate patient status in relation to the outcome, a five-point Likert scale to measure patient status, and a short list of references used in development of the outcome. The classification is also maintained by the Center for Nursing Classification and Clinical Effectiveness, the University of Iowa College of Nursing (Moorhead 2004).

This terminology may be used to follow patient outcomes throughout an illness episode or over an extended period of care. Each outcome has a unique code number that facilitates the use of outcomes in computerized clinical information systems and the manipulation of data to answer questions about healthcare quality and effectiveness. NOC is recognized by ANA's congress of Nursing Practice Steering Committee on Databases to support Clinical Nursing Practice as a classification system useful for clinical nursing practice. It has been included in NLM's UMLS (NAHIT 2005).

❖ CLINICAL CARE CLASSIFICATION

The Clinical Care Classification (CCC), formerly known as the Home Health Care Classification (HHCC), consists of two interrelated taxonomies; the CCC of Nursing Diagnoses and Outcomes and the CCC of Nursing Inventions and Actions. Each provides a standardized framework for documenting patient care in hospitals, home health agencies, ambulatory care clinics, and other healthcare settings. It consists of 21 care components that serve as a framework for mapping and linking the two interrelated CCC taxonomies to each other and to other health related classifications. They are used to track and measure patient/client care holistically over time, across settings, population groups, and geographic locations.

THE OMAHA SYSTEM

The Omaha System is a research based comprehensive taxonomy designed to generate data following usual or routine client care. It consists of three components (NAHIT 2005):

- Problem Classification Scheme
- Intervention Scheme
- Problem Rating Scale for Outcomes

It is a practice and documentation tool used by multi-disciplinary health care practitioners from the time of admission to discharge. The Omaha System translates data into information for practice, education, and research. EHR systems based on Omaha System provide an information management tool for data collection, aggregation, and analysis. Initial development and research began early in the 1970s. The most recent revision was published in 2005. The system is maintained by Martin Associates and a 12-member Omaha System Board (Martin 2005). The Omaha system was one of the first vocabularies recognized by ANA. It is also included in the NLM's UMLS Metathesaurus, the ANSI HISB Inventory of Clinical Information Standards, SNOMED International, LOINC, HL7, and in the accreditation standards of the JCAHO and the Community Health Accreditation Program (Martin 2005).

NURSING MANAGEMENT MINIMUM DATA SET

The Nursing Management Minimum Data Set (NMMDS) supports description, analysis, and comparisons of nursing care and nursing resources in light of the effects of context on complex healthcare outcomes. It was designed to complement the clinical patient oriented data designated in the Nursing Minimum Data Set (NMDS). (NAHIT 2005)

NURSING MINIMUM DATA SET (NMDS)

The NMDS provides uniform definitions and categories concerning the specific dimensions of nursing, which meet the information needs of multiple data users in the healthcare system. It was built on the concept of uniform minimum health data sets (UMHDSs) and includes the label and conceptual definition of the essential, specific elements that are used by nurses across all types of settings in the delivery of care. It also established comparability of nursing data across clinical populations, settings, geographical areas, and time. The NMDS was recognized by ANA in 1999 (NAHIT 2005).

The NMDS includes three broad categories of elements (NMDS 2003):

- Nursing care
- Patient or Client Demographics
- Service elements

PATIENT CARE DATA SET (PCDS)

The PCDS (version 4.0, 1998) consists of a three-component dictionary of

- Patient problems (363 terms)
- Patient care goals (311 terms)
- Patient care order (1,357 terms)

The intent of this set is to serve as a set of standard terms to represent and capture clinical data for inclusion in patient care information systems. It was recognized as one of the vocabularies considered for use by nurses and is included in the NLM's Metathesaurus. The PCDS are organized into 22 components (Ozbolt 1999):

- Activities
- Circulation
- Cognition
- Coping and Mental Health
- Fluids and Electrolytes

- Gastrointestinal Functions
- Health Knowledge and Behaviors
- Immunology
- Medications and Blood Products
- Metabolism
- Nutrition
- Physical Regulation
- Pre-, Intra-, and Post- Procedure
- Respiration
- Role Relationships
- Safety
- Self Concept
- Sensation, Pain, and Comfort
- Tissue Integrity
- Tissue Perfusion
- Urinary Elimination

INTERNATIONAL CLASSIFICATION FOR NURSING PRACTICE (ICNP)

The ICNP is a unified nursing language system into which existing into terminologies can be cross- mapped. It was created in 1989 by ICN. ICNP version 1 was scheduled for release in 2005(NAHIT 2005). It was established as a common language for describing nursing practice in order to improve communication among nurses, and between nurses and others. It can also be used to project trends in patient needs, provision of nursing treatments, resources, and outcomes of nursing care. ICNP is composed of the following elements;

- Nursing phenomena (nursing diagnoses)
- Nursing Actions (nursing interventions)
- Nursing Outcomes

The ICNP is maintained by the ICNP Program Director, University of Wisconsin-Milwaukee College of Nursing (ICNP2004).

UNIFIED MEDICAL LANGUAGE SYSTEM (UMLS)

The UMLS is a government-funded project from NLM. The UMLS has been in development in development since 1986. According to UMLS Fact Sheet, *"the purpose of UMLS is to aid the development of systems that help health professionals and researchers retrieve and integrate electronic biomedical information from a variety of sources and to make it easy for users to link disparate information systems, including computer-based patient records, bibliographic databases, factual databases, and expert systems"*. This goal is achieved through the three knowledge sources found in the UMLS:

1. The UMLS Metathesaurus
2. The SPECIALIST Lexicon, and
3. The UMLS Semantic Network.
   When looking in-depth at the UMLS, it is important to keep in mind that it has been designed for computer use; its layouts and so on are meant for machines.

❖ UMLS Metathesaurus

The UMLS Metathesaurus contains information on biomedical concepts and terms from more than 60 controlled vocabularies and classifications used in health records, administrative health data, bibliographic and full-text databases, and expert systems. It preserves the names, meanings, hierarchical contexts, attributes, and interterm relationships present in its source vocabularies; adds certain basic information to each concept; and establishes new relationship among terms from different source vocabularies. Its source vocabularies include terminologies designed for use in health records systems; large disease and procedure classifications used for statistical reporting and billing; more narrowly focused vocabularies used to record data related to psychiatry, nursing, medical devices, adverse drug reactions, and so on; disease finding terminologies from expert diagnostic systems; and some thesauri used in information retrieval.

Computer programs can use information in the Metathesaurus to interpret user inquiries, interact with users to refine their questions, identify the databases that contain information relevant to particular inquiries, and convert the user's terms into the vocabulary used by the relevant information sources. The scope of the Metathesaurus is determined by the combined scope of its source vocabularies.

The Metathesaurus is produced by the automated processing of machine-readable versions of its source vocabularies, followed by human review and editing by subject experts. It is intended primarily for use by system developers but also can be a useful reference tool for database builders, librarians, and other information professionals.

❖ UMLS SPECIALIST LEXICON

The UMLS SPECIALIST Lexicon is English – Language lexicon containing many biomedical terms. It has been developed in the context of the SPECIALIST natural language processing project at the NLM. The current version includes some 108,000 lexical records, with more than 186,000 strings (concept names).

The lexicon entry for each word or term record syntactic, morphological, and orthographical information (syntactic refers to formal properties of language; morphological refers to the study and description of word formation in a language, including inflection, derivation, and compounding; and orthographic refers to the correctness of spelling or the representation of the sounds of a language by written or printed symbols).

Lexical entries may be single-word or multi-word terms. Entries that share their base form and spelling variants, if any, are collected into a single lexical record. The base form is the inflected form of the lexical item – the singular form in the case of a noun, the infinitive form in the case of a verb, and the positive form in the case of an adjective or adverb.

❖ UMLS SEMANTIC NETWORK

The UMLS Semantic Network, through its 134 Semantic types, provides a consistent categorization of all concepts represented in the UMLS Metathesaurus. The 54 links between semantic types provide the structure for the network and represent important relationships in the biomedical domain. All information on specific concepts is found in the metathesaurus; the network provides information on the basic semantic types assigned to these concepts and defines their possible relationships.

NOSOLOGY

Nosology is the branch of medical science that deals with classification issues. Practically, Nosology is different from coding in that it is about all classification issues, whether they are related to reimbursement or other purposes.

As the need grows to formalize (classify in some manner) all health record data in any medium (text, image, voice) for purpose of outcomes research, decision support, and knowledge management, HIM professionals will have to expand their coding classification horizons to become *Nosologist*.

CHAPTER FIVE

INTERNATIONAL CLASSIFICATION OF DISEASES – 10TH REVISION (ICD10)

INTRODUCTION

The International Classification of Diseases and Related Health Problems, also known as ICD – 10 is the latest revision of the ICD published by the World Health Organization (WHO) in the year 1993. It is used to translate diagnoses of diseases and other health problems from words into alphanumeric codes, which permit easy storage, retrieval, and analysis of data.

The main innovation in this Tenth Revision was the use of an alphanumeric coding scheme of one letter followed by three numbers at the four-character level. This had the effect of more than doubling the size of the coding frame in comparison with the Ninth Revision, and enable the vast majority of chapters to be assigned a unique letter or group of letters, each capable of providing 100 three-character categories. Of the 26 available letters, 25 had been used, the letter 'U' being left vacant for future additions and changes and for possible interim classifications to solve difficulties arising at national and international level between revisions.

As a matter of policy, some three-character categories had been left vacant for future expansion and revision, the number varying according to the chapters; those with a primarily anatomical axis of classification had fewer vacant categories as it was considered that future changes in their content would be more limited in nature.

In this section, the basic design, structure, and presentation of the ICD – 10, with the several new innovations it contains shall be examined, in view of the improvements and advantages it possesses over the Ninth Revision of the ICD.

THE STRUCTURE OF THE ICD – 10

ICD – 10 comprises of three volumes:

Volume 1 – The Tabular List; contains the main classification in a hierarchical arrangement.

Volume 2 – Instructional Manual; provides guidance to users of the ICD, and

Volume 3 – The Alphabetical Index to the classification (vol.1)

VOLUME 1 – THE TABULAR LIST

Most of volume 1 is taken up with the main classification, composed of the list of three-character categories, tabular list of inclusions, and four-character subcategories. The 'core' classification – the of three-character categories – is the mandatory level for reporting to the WHO mortality database and for general international comparisons. This core classification also lists chapter and block titles. The tabular list, giving the full detail of the four-character level, is divided into 21 Chapters.

❖ THE CHAPTERS

The classification of ICD-10 is divided into 21 Chapters. The first character of the ICD code is a letter, and each letter is associated with a particular chapter, except for letter D, which is used in both chapter 2, Neoplasm, and chapter 3, Diseases of the blood and blood-forming organs and certain disorders involving the immune mechanism, and letter H, which is used in both chapter 7, Diseases of the eye and adnexa and chapter 8, Diseases of the ear and mastoid process. Four chapters (chapters 1,2,19, and 20) use more than one letter in the first position of their codes.

CHAPTER	DESCRIPTION	ALPHABETS
1	Infectious	A, B
2	Neoplasm	C,D
19	Injury & Poisoning	S,T
20	External causes	V,W,X,Y

For proper presentation, the following are the list of the 21 chapters of the ICD:

CHAPTER	DESCRIPTION	CODE RANGE
1.	Certain Infectious and Parasitic Diseases	A00-B99
2.	Neoplasm	C00-D48
3.	Diseases of Blood and Blood-Forming Organs, and certain disorders involving the immune mechanisms	D50-D89
4.	Endocrine, Nutritional, and Metabolic Diseases	E00-E90
5.	Mental and Behavioural Disorders	F00-F99
6.	Diseases of the Nervous System	G00-G99
7.	Diseases of the Eye and Adnexa	H00-H59

8.	Diseases of the Ear and Mastoid process	**H60-H95**
9.	Diseases of the Circulatory System	**I00-I99**
10.	Diseases of the Respiratory System	**J00-J99**
11.	Diseases of the Digestive System	**K00-K93**
12.	Diseases of the Skin and Subcutaneous Tissues	**L00-L99**
13.	Diseases of the Musculoskeletal System and Connective Tissues	**M00-M99**
14.	Diseases of the Genitourinary System	**N00-N99**
15.	Pregnancy, Childbirth, and the Puerperium	**O00-O99**
16.	Certain Conditions Originating in the Perinatal period	**P00-P96**
17.	Congenital Malformations, Deformations and Chromosomal Abnormalities	**Q00-Q99**
18.	Symptoms, Signs, and Abnormal Clinical and Laboratory Findings – Not Elsewhere Classified	**R00-R99**
19.	Injury, Poisoning, and Certain Other Consequences of External Causes	**S00-S99, T00-T99**
20.	External Causes of Morbidity and Mortality	**V01-V99, W00-W99, X00-X99, Y00-Y98**
21.	Factors Influencing Health Status and Contact with Health Services	**Z00-Z99**

Each chapter contains sufficient three-character categories to cover its content; not all available codes are used, allowing space for future revision and expansion. Chapter 1 – 17 relate to diseases and other morbid conditions, and chapter 19 to injuries, poisoning and certain other consequences of external causes. The remaining chapters complete the range of subject matter nowadays included in diagnostic data. Chapter 18 covers symptoms, signs and abnormal clinical and laboratory findings, not elsewhere classified. Chapter 20, External causes of morbidity and mortality, was traditionally used to classify causes of injury and poisoning, but, since the Ninth Revision, has also provided for any recorded external causes of diseases and other morbid conditions. Finally, chapter 21, Factors influencing health status and contact with health services, is intended for the classification of data explaining the reason for contact with healthcare services of a person not currently sick, or the circumstances in which the patient is receiving care at that particular time or otherwise having some bearing on that person's care.

❖ BLOCK OF CATEGORIES

The chapters are subdivided into homogenous 'blocks' of three-character categories. In chapter 1, the block titles reflect two axes of classification – mode of transmission and broad group of infecting organisms. In chapter 2, the first axis is the behavior of the neoplasm; within behavior, the axis is mainly by site, although a few three-character categories are provided for important morphological types (e.g. leukaemias, lymphomas, melanomas, mesotheliomas, kaposi's sarcoma). The range of categories is given in parentheses after each block title.

❖ THREE-CHARACTER CATEGORIES

Within each block, some of the three-character categories are for single conditions, selected because of their frequency, severity or susceptibility to public health intervention, while others are for groups of diseases with some common characteristics. There is usually provision for "other" conditions to be classified, allowing many different but rarer conditions, as well as "unspecified" conditions to be included.

❖ FOUR-CHARACTER SUBCATEGORIES

Although not mandatory for reporting at the international level, most of the three-character categories are subdivided by means of a fourth, numeric character after a decimal point, allowing up to ten subcategories. Where a three-character category is not subdivided, it is recommended that the letter "X" be used to fill the fourth position so that the codes are of standard length for data processing. The four-character subcategories are used in whatever way is most appropriate, identifying, for example, different sites or varieties if the three-character category is for a single disease, or individual diseases if the three-character category is for a group of conditions.

The fourth character .8 is generally used for "other" conditions belonging to the three-character category, and .9 is mostly used to convey the same meaning as the three-character category title, without adding any additional information.

HIERARCHICAL STRUCTURE OF THE TABULAR LIST

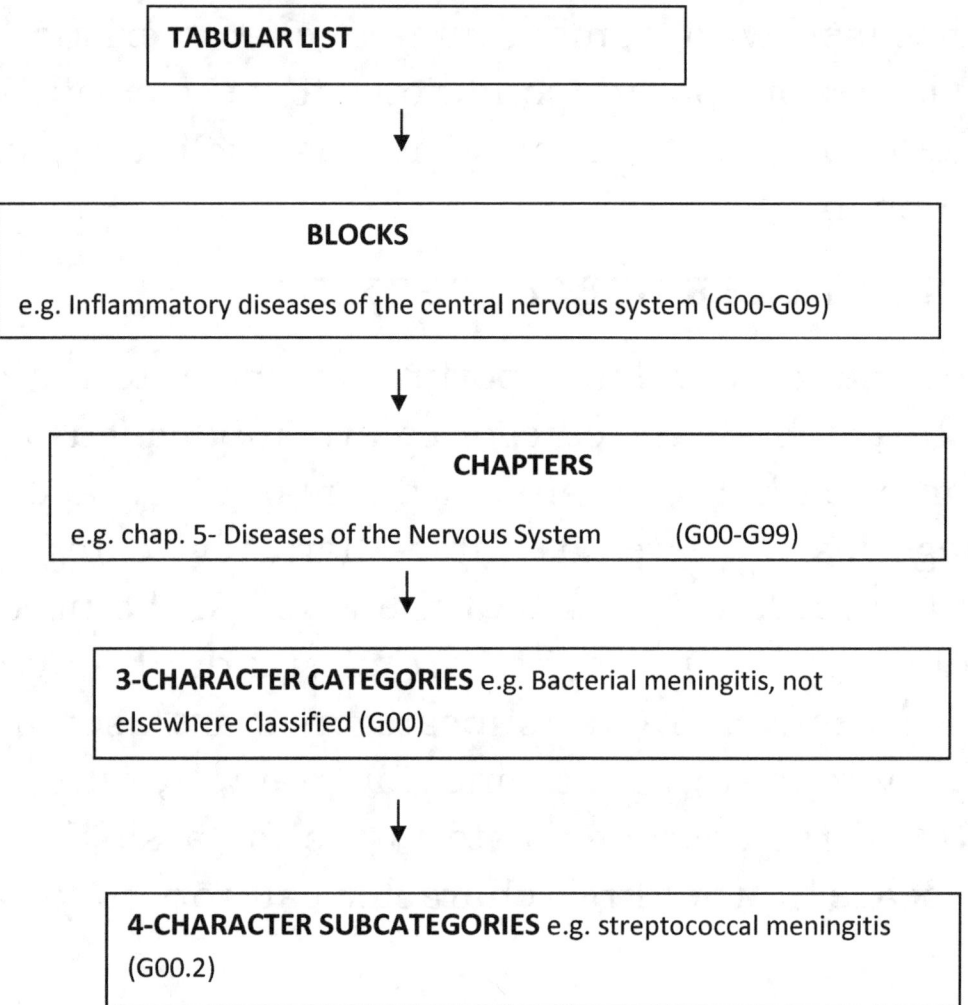

OTHER COMPONENTS OF VOLUME 1

❖ Morphology of Neoplasm – the classification of morphology of neoplasm may be used, if desired, as an additional code to classify the morphological type for neoplasm which, with a few exceptions, is classified in chapter 2 only according to behavior and site (topography). The morphology codes are the same as those used in the special adaptation of the ICD for Oncology (ICD – O).

❖ Special tabulation lists – because the full four-character lists of the ICD, and even the three-character lists, are too long to be presented in every statistical table, most routine statistics use a tabulation list that emphasizes certain single conditions and group others. The four special lists for the tabulation of mortality are an integral part of the ICD. Lists 1 and 2 are for general mortality, and lists 3 and 4 are for infant and child mortality (ages 0-4 years). There is also a special tabulation list for morbidity. Guidance on the appropriate use of the various levels of the classification and the tabulation lists is given in section 5 of this volume.

❖ Definitions – the definitions on pages 1233 – 1238 of volume 1 have been adopted by the World Health Assembly and are included to facilitate the international comparability of data.

❖ Nomenclature regulation – the regulations adopted by world Health Assembly set out the formal responsibilities of WHO member states regarding the classification of diseases and causes of death and the compilation and publication of statistics.

VOLUME 2 – INSTRUCTIONAL MANUAL

This volume of the Tenth Revision of the International Classification of Diseases and Related Health Problems (ICD-10) contains guidelines for recording and coding, together with much new material on practical aspects of the classification's use, as well as an outline of the historical background to the classification. This material is presented as a separate volume for ease of handling when reference needs to be made at the same time to the classification (volume 1) and the instructions for its use. Detailed instructions on the use of the Alphabetical Index are contained in the introduction to volume 3.

This manual provides a basic description of the ICD, together with practical instructions for mortality and morbidity coders, and guidelines for the presentation and interpretation of data. It is not intended to provide detailed training in the use of the ICD. The material included needs to be augmented with formal courses of instruction allowing extensive practice on sample records and discussion of problems.

SUMMARY OF THE CONTENT OF VOLUME 2 (INSTRUCTIONAL MANUAL)

Volume 2 contains:

- Description of the ICD-10 (pp. 1-17)
- How to use the ICD (pp. 18-29)
- Rules and guidelines for mortality and morbidity (pp. 30-123)
- Statistical presentation (pp. 124-138)
- History of the ICD development (pp. 139-150)

VOLUME 3 – THE ALPHABETICAL INDEX

The Alphabetical Index provides the index for the tabular list of the volume 1. Although the index reflects the provisions of the Tabular List in regard to the notes varying the assignment of a diagnostic term when it is reported with other conditions, or under particular circumstances (e.g. certain conditions complicating pregnancy), it is not possible to express all such variations in the index terms.

Volume 1 should therefore be regarded as the primary coding tool. The Alphabetical Index is, however, an essential adjunct to the Tabular list, since it contains a great number of diagnostic terms that do not appear in volume 1. The two volumes must therefore be used together.

The terms included in a category of the Tabular List are not exhaustive; they serve as examples of the content of the category or as indicators of its extent and limits. The Index on the other hand, is intended to include most of the diagnostic terms currently in use. Nevertheless, reference should always be made back to the Tabular List and its notes, as well as to the guidelines provided in volume 2, to ensure that the code given by the index fits with the information provided by a particular record.

ARRANGEMENT OF THE ALPHABETICAL INDEX

The Alphabetical Index is divided into three main sections as follows:

- Section 1 – contains all the terms classifiable to chapter 1-19 and 21, except drugs and other chemicals.
- Section 2 – contains the index of external causes of morbidity and mortality and contains all the terms classifiable to chapter 20, except drugs and other chemicals.
- Section 3 – contains the Table of Drugs and Chemicals, lists for each substance the codes for poisonings and adverse effects of drugs classifiable to chapter 19, and the chapter 20 codes that indicate whether the poisoning was accidental,

deliberate (self harm), undetermined, or an adverse effect of a correct substance properly administered.

SUMMARY OF THE CONTENT OF VOLUME 3 (ALPHABETICAL INDEX)

- Introduction (pp. 1-7)
- Alphabetical Index to diseases and nature of injury (pp. 9-572)
- External causes of morbidity and mortality (pp.573-624)
- Table of drugs and chemicals (pp. 625-746)
- Corrigenda to volume 1 (pp. 747-750)
- Lead terms are now printed in bold text, to improve readability of the index, and the table of neoplasm has had some changes.
- American spelling is used in volume 3.

STRUCTURE OF CODES IN ICD-10

- Major change is the move to alphanumeric codes.
- In ICD-10, the category code is made up of one alphabetic character followed by two numeric digits (as opposed to three numeric characters in ICD-9). For example,:

K37 (Acute appendicitis)

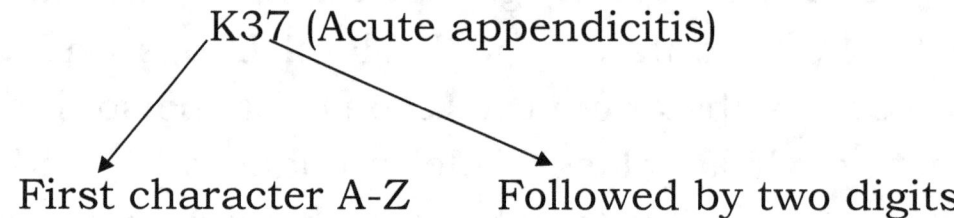

First character A-Z   Followed by two digits

Most of the three-character categories are further subdivided into subcategories to specify a disease or condition as specifically as possible. These subcategories are identified by adding a fourth digit. For example:

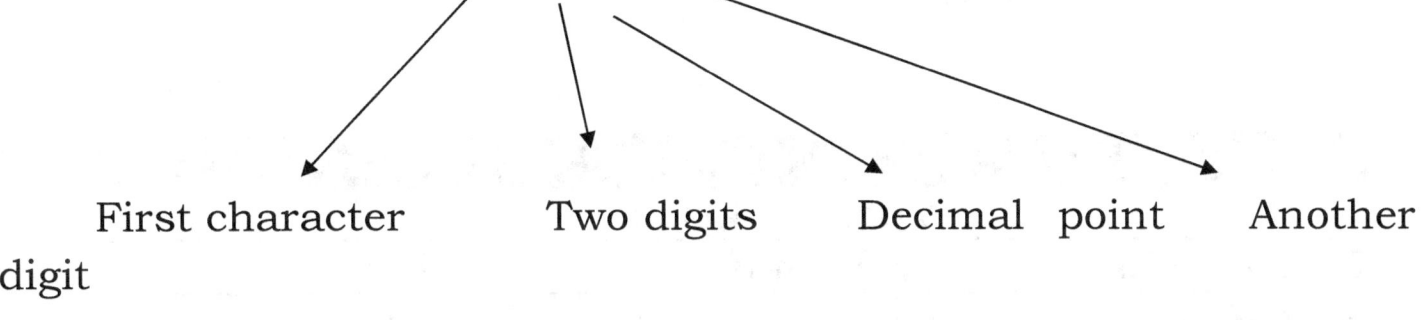

K 35 . 1

First character   Two digits   Decimal point   Another digit

VOLUME OF CODES

- By the use of alphabetical characters in ICD-10, it has provision to use 2600 categories. However, presently just over 2000 categories are used.
- Increase in codes is not uniform but rather concentrated in chapters relating to signs and symptoms, congenital abnormalities, and perinatal conditions
- Majority of codes in ICD-10 are four characters codes.

STRUCTURAL CHANGES IN ICD – 10

- ICD – 10 contains 3 volumes
- This was created by splitting;
    - a. Volume 1 of ICD-9 into volume 1 and volume 2.
    - b. Volume 2 of ICD-9 becomes volume 3.

REASSIGNMENTS

Conditions with a recently discovered etiology or new treatment protocol have been reassigned to a more appropriate chapter in ICD-10. These include;

CONDITION	ICD – 10	ICD – 9
Gout	Musculoskeletal	Endocrine
Bradycardia	Symptoms	Circulatory
Sarcodiosis	Blood	Infection
Refractory Anemia	Neoplasm	Blood

DIFFERENCES BETWEEN ICD – 9 AND ICD – 10

ICD – 9	ICD – 10
Introduced in 1977	Introduced in 1992
Entitled "International Statistical Classification of Diseases, Injuries and causes of Death"	Entitled "International Statistical Classification of Diseases and Related Health Problems"
Brought out in two volumes; **Vol. 1. Tabular List** **Vol. 2. Alphabetical Index**	Brought out in 3 volumes; Vol. 1. Tabular List Vol. 2. Instructional Manual Vol. 3. Alphabetical Index
Totally numeric in nature	Alphanumeric in nature
Has 17 chapters and 2 special supplementary classifications chapters	Has 21 chapters and no special supplementary chapters

Has "E" and "V" supplementary classifications	"E" and "V" supplementary classifications incorporated as part of the core classification
Includes the "Basic Tabulation List"	"Basic Tabulation List" withdrawn
Two short lists (each of 50 causes) introduced to list morbidity and mortality	These short lists withdrawn and replaced by 5 newly designed lists called "Special Tabulation Lists"
A separate classification, termed "Classification of Industrial Accidents, According to injury" introduced	This classification withdrawn altogether.
"Medical Certification and Rules for Classification" component as part of vol. 1, the tabular list.	This component removed from vol. 1 and features, in greatly revised form in vol. 2, the instructional manual.
Causes of Death component of the standard WHO Death Certificate has 3 lines (a, b, and c)	An additional (line "d") has been introduced to bring greater accuracy in determining the exact underlying cause of death.
Provides no guidelines about presenting statistical data in a standard format	Has a special chapter in volume 2 (Statistical Presentation) for this purpose
Alternative forms of diseases or condition indicated by identifying them under the main condition	Use of bullets to indicate alternative forms of disease or condition

"Diseases of the Nervous System and Sensory Organs" presented in chapter 5	Chapter 5 has been split into 3 chapters • Diseases of the Nervous System • Diseases of Eye and Adnexa • Diseases of Ear and Mastoid Process
No provision for post procedural disorders which often constitute a medical care problem in their own right	New categories are created at the end of certain chapters for post procedural disorders
Provisions of separate "Late Effect" category for each intent (i.e. suicide, accident, homicide, etc)	All such categories have been brought together in a block entitled " Sequelae of External causes of morbidity and mortality
"Type of Injury" was used as the axis of classification and then sub-classified the type of injury according to "site of injury"	Injuries are coded first to the "Body Region" where the injury has occurred, then to the "type of injury"
The main axis classification for land transport accidents was whether the event was a traffic or non-traffic accident	The main axis for this type of accident is now the injured person's "mode of transportation"
"Corrosion" and "Burns" were coded to the same set of codes as all other types of burns	Each 3-character category in the block identifies "corrosion" and "Burns" separately at the 4-character

	level.
"Friction Burns" were classified with "Superficial Injuries"	"Friction Burns are included with "Burns"
No "Activity Code" existed in the chapter for External Causes of Injury and Poisoning"	An "Activity Code" is provided in chapter 20 (External Causes of Morbidity and Mortality) for optional use to indicate the activity of the person at the time of the accident.
Both the "lead term" and "modifiers" in vol.2 (Index) were printed in usual normal size letters.	The lead terms in vol. 3 (Index) are printed in bold letters to improve the readability of the index
No provision for updating the ICD between Revisions – such activity only possible after every ten years	Provision for updating ICD between Revisions by issuing a small number of amendments annually.

INTRODUCTION TO ICD-11 (Review and Implementation)

According to the available information on the World Health Organization website (www.who.int), An external review of the ICD-11 Revision has been completed. The report notes the progress in the ICD Revision, and makes clear recommendations about forward progress in the revision.

WHO welcomes the constructive messages of the Report of the ICD-11 Revision Review. WHO is initiating the second phase of the revision process, acting immediately on the Review's recommendations.

In line with the updated recommendations, WHO has also updated the ICD-11 project plan and associated timelines.A Revision Steering Group (RSG) served as a consultative expert authority during the revision process from December 2007 through October 2016. The primary focus of the RSG was to provide guidance through reviewing the content to ensure adequate coverage of the full scope of health care diseases and related health conditions while addressing the needs of users.

The RSG is composed of the Co-Chairs of each content and classification Topic Advisory Group, as well as the co-chairs of the WHO-FIC Network and the WHO-FIC Committees and Reference Groups. The RSG is coordinated by a Small Executive Group known as the RSG-SEG, which includes the RSG Chair as well as six other individuals with broad areas of expertise drawn from the RSG.

Topic Advisory Groups served as the planning and coordinating advisory bodies for specific issues which are key topics in the update and revision process, namely Oncology, Mental Health, External Causes of Injury, Communicable Diseases, Non-communicable Diseases, Rare Diseases and others.

The primary charge of each group was to advise WHO in all steps leading to the revision of topic sections of ICD in line with the overall revision process. This included, in particular, advising on the development of various drafts of topic segments in line with the overall production timeline, reviewing and commenting on proposals from other stakeholders and experts while consolidating all input to achieve consistency across groups and areas, and many other tasks as outlined in the Terms of Reference for TAGs.

More than 200 scientists and other experts from more than 35 countries and all WHO Regions have contributed to the work.

The current ICD Revision structures are being reorganized towards a long term maintenance framework. This includes revisiting the governance design, and evolving the status quo into a new proposed structure. Among other new structures, WHO will launch a Medical Scientific Advisory Committee (MSAC) at the Revision Conference in 2016 comprised of approximately 6-10 experts selected by WHO. The main role of the MSAC will be to advise on scientific content for the ICD-11.

The MSAC will review all proposals in parallel and will be consulted on medical and scientific questions arising from the Network, as well. The MSAC is also responsible for providing advice on medical and scientific information in the foundation. Additional Special Projects may be established to develop and evaluate links to other classifications and terminologies and to advise on the associated informatics and architecture considerations. ITC may play a role in these projects.

PROPOSED CHAPTERS OF THE ICD-11

01 Certain infectious or parasitic diseases

02 Neoplasms

03 Diseases of the blood or blood-forming organs

04 Diseases of the immune system

05 Endocrine, nutritional or metabolic diseases

06 Mental, behavioural or neurodevelopmental disorders

07 Sleep-wake disorders

08 Diseases of the nervous system

09 Diseases of the visual system

10 Diseases of the ear or mastoid process

11 Diseases of the circulatory system

12 Diseases of the respiratory system

13 Diseases of the digestive system

14 Diseases of the skin

15 Diseases of the musculoskeletal system or connective tissue

16 Diseases of the genitourinary system

17 Conditions related to sexual health

18 Pregnancy, childbirth or the puerperium

19 Certain conditions originating in the perinatal period

20 Developmental anomalies

21 Symptoms, signs or clinical findings, not elsewhere classified

22 Injury, poisoning or certain other consequences of external causes

23 External causes of morbidity or mortality

24 Factors influencing health status or contact with health services

25 Codes for special purposes

26 Traditional Medicine conditions - Module I

V Supplementary section for functioning

X Extension Codes

CHAPTER SIX

CLINICAL CODING AND INDEXING

INTRODUCTION

Traditionally, Health Information Management (HIM) professionals have been responsible for the process of coding and indexing, or clinical code assignment, in the healthcare organizations. Historically and factually, the coding function has served two primary purposes:

I.   *To generate statistics and create secondary records (for example, diagnostic and procedural or surgical indexes) and,*

II. *To create documentation for reimbursement purposes.*

In the light of the above purposes emanating from coding functions, particularly the second purpose on reimbursement, any healthcare system that trifles with issues of coding is in a serious peril. Why? In any healthcare system, where the majority of expenses reimbursed by third-party payer such as Managed Care and Insurance Companies rather than the patients themselves, the only key for an effective reimbursement and payment system is in coding. For example, in the United States of America (USA) healthcare system, where a third party payer system works, such a system cannot joke with the coding functions of the HIM. As healthcare costs have risen over the past thirty years, third-party payers have searched for ways to contain costs. With the implementation of Prospective Payment System based on coded clinical data, procedural reviews of documentation to substantiate code assignment have become routine. In order words, accurate coding has become central to the financial survival of healthcare provider organizations.

However, a healthcare system in a developing nation like Nigeria can take a cue from this, especially with the just implemented National Health Insurance Scheme (NHIS), which is a semblance of the of the U.S. healthcare system. If the NHIS is to survive, and for it to serve the purpose of making healthcare available to all, which is key to the drive of the country to attain the Millennium Development Goals (MDGs). For healthcare organizations not to run at a loss, serious attention must be paid to the clinical coding functions.

For the HIM professionals involved in coding, their first responsibility is to ensure the accuracy of coded data. To this end, there must be strict adherence to the established code of professional ethics by which coders must abide. Obviously, there is no separate code of ethics for coders in Nigeria presently, but this could be established[1]. The absence of such ethical code is an eloquent evidence of the little or no importance attached to coding by policy makers even the healthcare provider institutions themselves. The general code of ethics for HIM professional in Nigeria may not adequately spell out what is expected of coders, in view of the importance of coding.

This chapter examines the entirety of clinical coding and indexing functions in the contemporary times, as it is obtained in the developed world, thereby making it easy for practitioners in Nigeria and other developing states to learn and borrow a leaf.

WHAT IS CLINICAL CODING?

---

[1] AHIMA code of ethics for coders is included elsewhere in this books

Generally, clinical coding is defined as the process of assigning numeric representations to clinical documentation

In the context of ICD-10, clinical coding can be defined as the process of assigning alphanumeric representation to clinical documentation.

Whichever way it goes, the fact remains that coding helps to reduce and compress every diagnostic and/or procedural documentation about a patient into a uniform coded data. By joining the above definitions together, a proper way of defining clinical coding should read thus:

'The process of assigning numeric and/or alphanumeric representations to clinical documentation (i.e. specific diseases, diagnoses and/or procedures) according to a classification system (i.e. ICD-10, ICD-10-CM, ICD-10-PCS, ICD-O, etc)'.

From the above definition, the clinical documentations to be coded are usually in the form of a disease, diagnosis, or procedure. For the purpose of clarity, it is pertinent to look more deeply into the meaning of a diagnosis:

A diagnosis is made on the basis of extensive knowledge about the patient such as family history, physical examination and investigation including X-rays and laboratory tests. There different variants of diagnosis:

- ***Clinical Diagnosis*** – this is based upon symptoms shown during life, irrespective of the morbid changes producing them.

- **Principal Diagnosis** – the condition established after careful study, to be chiefly responsible for occasioning the admission of the patient to the hospital for care.
- **Pathological Diagnosis** – this is based on gross and microscopic examination of the structural lesions present.
- **Differential Diagnosis** – this is based on symptoms and physical signs of two contrasting diseases
- **Provisional Diagnosis or Tentative Diagnosis** – this is based upon the availability of sources of information but subject to changes.
- **Pre-operative Diagnosis** – this is made before operation and based on clinical findings
- **Post-operative Diagnosis** – this is based upon findings observed during and after the operation.
- **Other Diagnosis** – all conditions which co-exist at the time of admission or develop subsequently and which affect the treatment received and/or the Length of Stay (LOS).

WHO IS A CLINICAL CODER?

A **clinical coder** – also known as **clinical coding officer**, **diagnostic coder**, **medical coder** or **medical records technician**– is a health information professional whose main duties are to analyse clinical statements and assign standard codes using a classification system. The data produced are an integral part of health information management, and are used by local and national governments, private healthcare organizations and international agencies for various purposes, including:

- medical and health services research,
- epidemiological studies,
- health resource allocation,
- case mix management,
- public health programming,
- medical billing, and
- public education.

For example, a clinical coder may use a set of published codes on medical diagnoses and procedures, such as the International Classification of Diseases (ICD) or the Common Coding System for Healthcare Procedures (HCPCS), for reporting to the health insurance provider of the recipient of the care. The use of standard codes allows insurance providers to map equivalencies across different service providers who may use different terminologies or abbreviations in their written claims forms, and be used to justify reimbursement of fees and expenses. The codes may cover topics related to diagnoses, procedures, pharmaceuticals or topography. The medical notes may also be divided into specialties for example cardiology, gastroenterology, nephrology, neurology , pul monology or orthopedic care.

A clinical coder therefore requires a good knowledge of medical terminology, anatomy and physiology, a basic knowledge of clinical procedures and diseases and injuries and other conditions, medical illustrations, clinical documentation (such as medical or surgical reports and patient charts), legal and ethical aspects of health information, health data standards, classification conventions, and computer- or paper-based data management, usually as obtained through formal education and/or on-the-job training.

The basic task of a clinical coder is to classify medical and health care concepts using a standardized classification. Most clinical coders are employed in coding inpatientepisodes of care. However, mortality events, outpatient episodes, general practitioner visits and population health studies can all be coded. Clinical coding has three key phases:
a) Abstraction
b) Assignment and
c) Review.

Abstraction
The abstraction phase involves reading the entire record of the health encounter and analyzing the information to determine what condition(s) the patient had, what caused it and how it was treated. The information comes from a variety of sources within the medical record, such as clinical notes, laboratory and radiology results, and operation notes.

Assignment
The assignment phase has two parts: finding the appropriate code(s) from the classification for the abstraction; and entering the code into the system being used to collect the coded data.

Review

Reviewing the code set produced from the assignment phase is very important. Clinical coder must ask themselves, "does this code set fairly represent what happened to this patient in this health encounter at this facility?" By doing this, clinical coders are checking that they have covered everything that they must, but not used extraneous codes. For health encounters that are funded through a case mix mechanism, the clinical coder will also review the diagnosis-related group (DRG) to ensure that it does fairly represent the health encounter.

COMPETENCY LEVELS

Clinical coders may have different competency levels depending on the specific tasks and employment setting.

Entry-level / trainee coder

An entry level coder has completed (or nearly completed) an introductory training program in using clinical classifications. Depending on the country; this program may be in the form of a certificate, or even a degree; which has to be earned before the trainee is allowed to start coding. All trainee coders will have some form of continuous, on-the-job training; often being overseen by a more senior coder.

Intermediate level coder

An intermediate level coder has acquired the skills necessary to code many cases independently. Coders at this level are also able to code cases with incomplete information. They have a good understanding of anatomy and physiology along with disease processes. Intermediate level coders have their work audited periodically by an Advanced coder.

Advanced level / senior coder

Advanced level and senior coders are authorized to code all cases including the most complex. Advanced coders will usually be credentialed and will have several years of experience. An advanced coder is also able to train entry-level coders.

Nosologist

A nosologist understands how the classification is underpinned. Nosologists consult nationally and internationally to resolve issues in the classification and are viewed as experts who can not only code, but design and deliver education, assist in the development of the classification and the rules for using it.

Nosologists are usually expert in more than one classification, including morbidity, mortality and casemix. In some countries the term "nosologist" is used as a catch-all term for all levels.

PRINCIPLES AND RULES OF CODING *(USING ICD-10 CLASSIFICATION SYSTEM)*

I. The Alphabetical index is utilized to locate the main entry term. The arrangement of the index is by conditions in the disease index

II. Conditions may be expressed as adjectives, nouns or eponyms in the alphabetical index of terms. Some conditions have more than one listing and may be located under either one.

III. To select the appropriate code, read and be guided by any notes that appears under the main term:

a. *Terms enclosed in parentheses following the main entry as well as sub terms under the main entry;*

b. *Appropriate sites or modifiers listed in alphabetical sequence under such main terms with further sub terms listings as necessary;*

c. *Eponyms appearing as both main term entries and modifiers under such main terms as diseases or syndromes;*

d. *Conditions expressed as adjectives appearing in the list of main terms;*

e. *Cross references to synonyms, closely related terms and code categories beginning with 'see' and 'see also'.*

IV. Read and be guided by the list of inclusions or exclusions that may appear not only under the particular code, but also under the category code or section title for that particular code. Never code directly from the alphabetical index, for important instructions often appear in the tabular list. Watch for exclusion codes

V. Remember these special point:

b. Follow the instructions to 'code also', 'code also underlying disease' or 'use additional code if desired', whenever it appears, in order that all the component elements of a complex diagnostic statement or procedure may be fully identified. Two or more codes may be necessary.

c. The coder should continue coding diagnostic statement until all of the component elements of a complex diagnostic statement or operation are fully identified.

d. In some instances it may be necessary to ask the physician to add additional diagnoses when the content of the medical record suggests that the listing of diagnostic statement is incomplete.

e. Only diagnoses which related to the current episode of care need to be coded. Medical records often contain statements like 'status post-hysterectomy' or 'history of congenital heart disease' which shows no relationship to the current episode of care. A history of past conditions not relevant to the current episode of care should be submitted for coding purposes.

f. If an inpatient diagnosis is stated as 'suspected', 'questionable', 'likely', or in other similar terms, code the condition as if it existed or was established.

g. Inconclusive diagnosis expressed in terms of differential conditions should be coded as follows:

- *Comparative and contrasting diseases or conditions should be coded as being a suspected condition. Example: Acute Pancreatitis vs. acute Cholecystitis – use K85 and K81.*
- *Comparative and contrasting etiologies should be coded to disease or condition, cause not otherwise specified.*

> *Example; Acute Peritonitis, bile or generalized : use K65.9*
>
> - *Symptom followed by contrasting and comparative diagnosis should be coded with the symptom as the principal as the principal diagnosis code. All of the contrasting diagnoses should be coded as suspected conditions. Example; Fatigue due to either depressive reaction or hypothyroidism – use R53, F32.9 and F39 (in this sequence).*

h. The term 'rule out', 'ruled out', 'suspected', or 'probable' do not appear in the alphabetical index. However , there are rules to govern the coding of diagnosis is to be coded, as if it exists because the physician feels that there is a strong probability that it does not exist. But when the diagnosis is modified by the term 'ruled out', it is not coded because the physician has eliminated this possibility as a reason for the patient's symptoms. In this instance, the patient's symptoms would be coded. Example: Chest pain, rule out myocardial infarction. In this case, code the myocardial infarction rule out. In this case, code Chest pain R07.4.

i. When a specific condition is stated as both acute (or sub-acute) and chronic, and the alphabetical index provides codes at the third, fourth, or fifth digit level for acute and for chronic, use both codes. Examples: Acute and chronic non-rheumatic pericarditis – use   30.9 and   31.9.

STEPS TO CODING AND INDEXING (MANUAL)

Although the coding process may vary from organization to organization, or even within an organization between care settings, certain steps are usually followed when codes are assigned either manually or electronically. The following are the steps to take when coding and indexing is being done manually –

a. Receive completed records from the incomplete records control unit and acknowledge the receipt of them.

b. Code these records for disease classification as per ICD within the guidelines established by the department.

c. Code records for operation as per the ICSO within the guidelines established by the department.

d. Index the coded disease on the index card.

e. Index the coded operation on the operation index card.

f. Complete legibly with appropriate information all of the columns of the index cards.

g. Make cross reference entries on related index cards for those patients who have more than one disease.

h. Use asterisks marks in the result column for all secondary diagnoses

i. Type the name of the hospital, the code number, the name of disease or operation, the year and the card number before the patient's name is entered on the card.

j. Use blue ink for all discharged alive cases and red ink for all expired cases.

k. Place a checkmark by the code number in the file when it is entered on the index card.

l. Verify all records to ensure that they have been coded and indexed

m.   Place a checkmark by the appropriate number in the coding and indexing control register.

n. Forward the coded and indexed records to the incomplete records control unit and obtain their acknowledgement as a token of having received the records.

APPLICATION OF TECHNOLOGY TO THE CODING PROCESS

Recent technologies are having considerable impact on the coding process, and new technologies hold promise for the future. Some of the most significant advances are discussed below:

ENCODERS

An encoder is a computer software program designed to assist coders in assigning appropriate clinical codes to words and phrases expressed in natural human language (Slee and Slee 1991). Initially developed during the era of paper-based health records, the principal purpose of encoder was to help ensure accurate reporting for reimbursement (Beinborn 1999).

Encoder come in two distinct categories: *logic-based and automated codebook formats.*

A logic-based encoder prompts the user through a variety of questions and choices based on the clinical terminology entered. The coder selects the most accurate code for a service or condition (and any possible complications or co-morbidities).

An automated codebook provides screen views that resemble the actual format of the coding system. This allows the coder to review code selections, notes, look-up tables, edits, and various other automated notations that help him or her to choose the most accurate code for a condition or service. Although, encoders can cite official coding guidelines and provide code optimization guidance, they require user interaction. Encoders promote accuracy as well as consistency in the coding of diagnoses and procedures.

AUTOMATED CODE ASSIGNMENT

Rather than prompting the user to make various selections on the basis of patient record documentation, automated code assignment uses data that have been entered into the computer to automatically assign codes (Beinborn 1999). Automated code assignment uses Natural Language Processing (NLP) technology to read the data contained in a Computerized Patient Record (CRP). The Natural Language Processing technology used might be algorithmic (rule based) or statistical. Schnitzer (2000) has stated that statistical approaches that can predict how an experienced coder might code a record and use machine learning techniques are superior to rigid rule sets. This opinion is congruent with the oft-heard argument that "coding is subjective"

According to Schnitzer, the advantages to using automated code assignment include:

- **Consistency** – *automated code assignment is consistently correct and consistently incorrect; thus, errors, once detected, are easier to locate and fix.*
- **Accuracy** – *as with medical knowledge, the amount of information that must be synthesized to correctly code patient records has increased substantially. Expert computer software programs can apply these rules and guidelines accurately.*
- **Speed** – *the computational power of computers can apply the many coding rules and guidelines efficiently.*

It is important to note, however, that automated code assignment is not a magic bullet. For example, it cannot address the major obstacle facing today's human coder; the lack of accurate, complete clinical documentation. As Schnitzer   (2000) put it, "If a service isn't appropriately documented, there is no way for NLP to find it, assume it, or infer it". In addition, automated code assignment has not developed to the point where it can operate without substantial human interaction (Warner 2000). Finally, the most effective automated code assignment system will need to accommodate any weaknesses and deficiencies in the classification and coding systems it uses (Beinborn 1999).

As automated code assignment continues to evolve, so too will the role of the coding professional. It was a vision actually that coding professional would metamorphosed to clinical data specialist in the future, and it was also predicted that this individuals will manage data in a number of areas, including clinical coding, outcome measurement, specialty registries, and research databases. Beinborn (1999) stated that coders will be essential to the implementation and maintenance of automated code assignment systems as well as of decision support systems, data interpretation, and other evolving areas that rely on healthcare data.

EMERGING TECHNOLOGIES

As is apparent from the preceding subsection, the status quo will not continue for coding. A number of emerging technologies will likely be used to support the coding function.

Speech recognition is one candidate for improving the coding process as well as coding accuracy. As speech recognition improves and has the ability to accurately document what the clinician is saying, the case of completely documenting healthcare services also improves. Schwager (2000, p. 64) has maintained that "speech recognition, combined with technology designed to extract and structure medical information (Natural Language Processing) contained in narrative text, can automate the coding process used in reimbursement". He also has discussed the possibility that the computerized patient record software could analyze the record in an interactive fashion, prompting the clinician for higher-quality documentation.

Hand held Computers or Personal Digital Assistants (PDAs) also are becoming commonplace. Many physicians and other healthcare professionals find them helpful. As computer systems continue to develop, it will only be a matter of time before these small units are used for data entry and other functions in healthcare organizations.

CODING QUALITY CONTROL

The quality of coded clinical data depends on a number of factors including:

- Adequate training for everyone involved in the coding process, including coders, coding supervisors, clinicians and financial personnel.
- Adequate references and support resources, including up to date coding books, as well as subscription publications that communicate official and practice guidelines, and in some cases, encoders and support software.
- Accurate and complete clinical documentation that includes every pertinent condition and service provided to the patient.
- The support of top management, who must understand how important the coding function is to the organization's continued existence.
- A performance improvement plan for the coding function that ensures continuous quality improvement processes.
- When all of these components are present, the quality of coded clinical data can be evaluated according to the following elements:

    I. **Reliability:** *the extent to which data can be reproduced by subsequent measurements or tests (for example, coded clinical data are considered reliable when multiple coders assign the same codes to a record)*

    II. **Validity:** *the extent to which coded data accurately reflect the patient's diagnoses and procedures (Bowman 2001, p. 250)*

    III. **Completeness:** *the extent to which the coded data represent all of the patient's relevant diagnoses or procedures.*

*IV. **Timeliness:*** *the extent to which the coded data are available within the time frames required for billing purposes, decision support, and other uses.*

CODING POLICIES AND PROCEDURES

As with other organizational policies and procedures, coding policies and procedures are needed to promote consistency. Items to be included in coding policies and procedures include:

- Directions for reviewing the record.
- Instructions on how to address incomplete or conflicting documentation
- Instructions for communicating with physicians and developing physician queries and for clarification and recording health record addenda.
- Instructions on the actions to be taken when an appropriate code cannot be located.
- Use of codes not required for reimbursement (optional codes).
- Standardized definitions or code sets (for example, WHO requirements)
- Use of reference materials and books and instructions for updating
- Computerized data entry or other processes (Bowman 2001, p. 251)

QUALITY ASSESSMENT FOR THE CODING PROCESS

Quality assessment for the coding process is also referred to as performance improvement for the coding process. It involves looking at more than just the assigned codes. As stated previously, accurate coding is essential to the economic survival of the healthcare organization; hence, ongoing efforts to improve the coding process should better economic benefits.

Most important to improving any coding processes is to first document and understand the current coding process. The preceding section can be used as a guide to documenting that process, remembering that the actual process is different in every organization. One of the most important items in the coding process is the quality of the clinical documentation. Even though HIM professionals have little or no direct control over the documentation, they should work with the clinicians and other personnel to continue to improve its quality.

After the current coding process has been documented, quality measurement points should be established at specific stages in the process. Baseline measurements need to be taken, and benchmarks for improvement should be set up.

Quality assessment of a process is generally ongoing. Even when the most ambitious goals for the most important measures have been accomplished, the measured should continue to be monitored. In addition, management should continually search for ways to improve the coding process through computerization or other methods.

DISCHARGE SUMMARY

The *discharge summary*, also called the **clinical résumé**, provides details about the patient's stay in the facility and is the foundation for future treatment. It is prepared when the patient expires. The summary states the patient's reason for admission and gives a brief history explaining why he or she needed to be hospitalized. Pertinent laboratory, X-ray, consultation, and other significant findings; as well as the patient's response to treatment to treatment or procedures, are included. In addition to a description of the patient's condition at discharge, the discharge summary delineates specific instructions given to the family for future care, including information on medications, referrals to other providers, diet, activities, follow-up visits to the physician, and the patient's final diagnoses. The discharge summary must be signed and dated by the physician.

Some facilities require the final diagnoses to be recorded and the discharge summary to be dictated or written at discharge or they immediately consider the chart delinquent. The information in the discharge is extremely important to meet the facility's needs of coding, billing and reimbursement. In some facilities, a paper discharge summary form with an outline of contents is used to ensure that all items are included. The form has three parts so that the part with follow-up instructions can be given to the patient.

When a patient dies in the hospital, hospitals often require the physician who pronounced death to write a note that gives the time and date of death. The death note is in addition to the discharge summary required in all death cases no matter how long the patient was in the facility. Most facilities do not require a discharge summary for normal newborns and obstetrical cases without complications, as long as there is a final progress note.

A discharge summary is not a required for patient who are in the hospital for 48 hours or less. Such patients usually have a short – stay or short – service record or a final discharge progress note. This one page form can be used to record the history and physical examination, the operative report, the discharge summary, and discharge instructions. When the patient is admitted to the facility, the reason for admission must be recorded. When the patient dies 48 hours or less after admission, the short – stay record is insufficient and a complete discharge summary must be prepared.

The discharge summary must be completed within 30 days after the date of discharge. When a patient is transferred, the physician should complete the discharge summary within 24 hours.

The principal diagnosis and other diagnoses should be recorded completely without symbols or abbreviations on the health record summary sheet (the face sheet) or discharge summary or on another form prescribed by the facility. The principal diagnosis is defined as the condition determined, after study, to be chiefly responsible for occasioning the patient's admission to the hospital.

Healthcare facilities must determine what information goes with the patient when he or she is transferred to another level of care. When the transfer is to an affiliated institution that is part of the same healthcare system, the original patient record is transferred with the patient and new orders are written at the receiving institution to initiate care. A discharge summary is generally required.

CHAPTER SEVEN

UNDERSTANDING THE USE OF THE ICD-10

INTRODUCTION

All users of the ICD must be aware of certain practical information that would facilitate good knowledge and easy understanding of the purpose and structure of the ICD. This is very vital for statisticians and analysts of health information as well as for coders. Accurate and consistent use of the ICD depends on the correct application of all three volumes.

This section contains certain practical information which all users need to know in order to exploit the classification to its full advantage. Included in this chapter are some specific technical terms and established conventions used in the ICD-10.

INCLUSION TERMS

Within the three- and four-character rubrics[2], there are usually listed a number of other diagnostic terms. These are known as "inclusion terms" and are given, in addition to the title, as examples of the diagnostic statements to be classified to that rubric. They may refer to different conditions or be synonyms. They are not a sub-classification of the rubric.

Inclusion terms are listed primarily as a guide to the content of the rubrics. Many of the items listed relate to important or common

---

[2] In the context of the ICD, "rubric" means either a three-character category or a four-character subcategory.

terms belonging to the rubric. Others are borderline conditions or sites listed to distinguish the boundary between one sub-category and another. The lists of inclusion terms are by no means exhaustive and alternative names of diagnostic entities are included in the Alphabetical Index, which should be referred to first when coding a given diagnostic statement.

EXCLUSION TERMS

Certain rubrics contain lists of conditions preceded by the word "Excludes". These are terms which although the rubric title might suggest that they were to be classified there, are in fact classified elsewhere. An example of this is in category A46, "Erysipelas", where postpartum or puerperal erysipelas is excluded. Following each excluded term, in parentheses, is the category or subcategory code elsewhere in the classification to which the excluded term should be allocated.

General exclusions for a range of categories or for all subcategories in a three-character category are to be found in notes headed "Excludes", immediately following a chapter, block or category title.

GLOSSARY DESCRIPTION

In addition to inclusion and exclusion terms, Chapter 5, Mental and Behavioral Disorders, uses glossary descriptions to indicate the content of rubrics. This device is used because the terminology of mental disorders varies greatly, particularly between different countries, and the same name may be used to describe quite different conditions. The glossary is not intended for use by coding staff.

Similar types of definition are given elsewhere in the ICD, for example, Chapter 21, to clarify the intended content of a rubric.

THE "DAGGER AND ASTERISK" SYSTEM (†/*)

This system of "dagger and asterisk" was first introduced in ICD-9, and it continued in ICD-10. This is a system whereby there are two codes for diagnostic statements containing information about both an underlying generalized disease and a manifestation in a particular organ or site which is a clinical problem in its own right.

The primary code is for the underlying disease and is marked with a dagger (   ); an optional additional code for the manifestation is marked with an asterisk (*). This convention was provided because coding to underlying disease alone was often unsatisfactory for compiling statistics relating to particular specialties, where there was a desire to see the condition classified to the relevant chapter for the manifestation when it was the reason for medical care.

While the dagger and asterisk system provides alternative classifications for the presentation of statistics, it is a principle of the ICD that the dagger code is the primary code and must always be used. Provision should be made for the asterisk code to be used in addition if the alternative method of presentation may also be required. For coding, the asterisk code must never be used alone. Statistics incorporating the dagger code conform to the traditional classification for presenting data on mortality and morbidity and other aspects of medical care.

Asterisk codes appear as three-categories. There are separate categories for the same conditions occurring when a particular disease is not specified as the underlying cause. For example, categories G20 and G21 are for forms of Parkinsonism that are not manifestations of other diseases assigned elsewhere, while G22* is for "Parkinsonism in diseases classified elsewhere". Corresponding dagger codes are given for conditions mentioned in asterisk categories; for example, for syphilitic Parkinsonism in G22*, the dagger code is A52.1[†]   .

ESTABLISHED CONVENTIONS USED IN THE ICD

Generally, conventions have been described as "traditional methods or styles used in presenting technical information, which if unexplained may be undecipherable to a lay person". There are a good number of such conventions used in the ICD. They are hereby explained as they appear in the respective volumes of the ICD.

CONVENTIONS USED IN THE TABULAR LIST (VOLUME 1)

In listing inclusion and exclusion terms in the tabular list, the ICD employs some special conventions relating to the use of parentheses, square brackets, braces, colons, the abbreviation "NOS", the phrase "not elsewhere classified" (NEC) and word "and" in titles. These need to be clearly understood by coders and by anyone wishing to interpret statistics based on the ICD.

PARENTHESES (   )

Parentheses are used in volume 1 in four important situations:

a)   Parentheses are used to enclose supplementary words, which may follow a diagnostic term without affecting the code number to which the word outside the parentheses would be assigned. For example, in I10 the inclusion term "Hypertension (arterial) (benign) (essential) (malignant) (primary) (systemic)", implies that I10 is the code number for the word "Hypertension" alone or when qualified by any, or any combination, if the words in parentheses.

b)   Parentheses are also used to enclose the code to which an exclusion term refers.   For   example;   H01.0,   Blepharitis,   excludes blepharoconjunctivitis (H10.5)

c)   Another use of parentheses is in the block titles, to enclose the three-character codes of categories included in that block.

d)   The last use of parentheses was incorporated in the Ninth Revision and is related to the dagger and asterisk system. Parentheses are used to enclose the dagger code in an asterisk category or the asterisk code following a dagger term.

SQUARE BRACKET [   ]

The square brackets are used:

For enclosing synonyms, alternative words, or explanatory phrases, for example, A30 Leprosy [Hansen's disease];

For referring to previous notes; for example, C00.8 Overlapping lesion of lip [see note 5 on p. 182];

For referring to a previously stated set of fourth character subdivisions common to a number of categories; for example, K27 Peptic ulcer, site unspecified [see page 566 for subdivisions].

COLON :

A colon is used in listing of inclusion and exclusion terms when the words that precede it are not complete terms for assignment to that rubric. They require one or more of the modifying or qualifying words indented under them before they can be assigned to the rubric. For example, in K36, "other appendicitis" , the diagnosis "appendicitis" is to be classified there only if qualified by the word "chronic" or "recurrent".

BRACE   }

A brace is used in listing of inclusion and exclusion terms to indicate that neither the words that precede it nor after it are complete terms. Any of the terms before the brace should be qualified by one or more of the terms that follow it. For example:

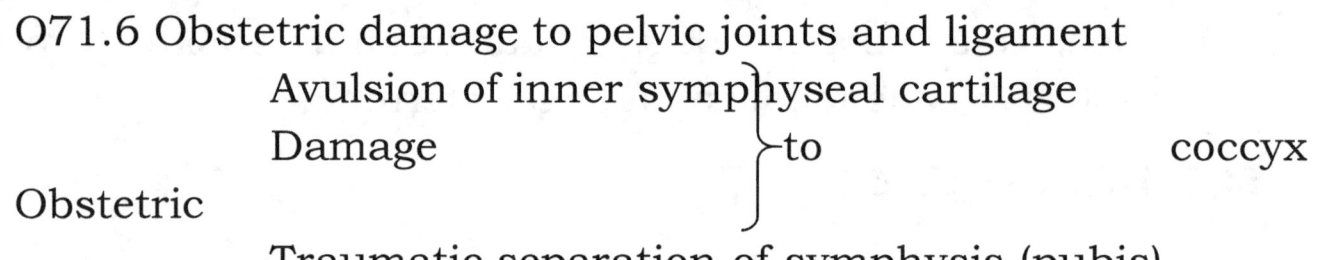

O71.6 Obstetric damage to pelvic joints and ligament
       Avulsion of inner symphyseal cartilage
       Damage   }to   coccyx
 Obstetric
       Traumatic separation of symphysis (pubis)

"NOS" (NOT OTHERWISE SPECIFIED)

The letters NOS are an abbreviation for "Not Otherwise Specified", implying "unspecified" or "unqualified".

Sometimes an unqualified term is nevertheless classified to a rubric for a more specified type of the condition. This is because, in medical terminology, the most common form of a condition is often known by name of the condition itself and only the less common types are qualified. For example, "mitral stenosis" is commonly used to mean "rheumatic mitral stenosis". These inbuilt assumptions have to be taken into account in order to avoid incorrect classification. Careful inspection of inclusion terms will reveal where an assumption of cause has been made; coders should be careful not to code a term as unqualified unless it is quite clear that no information is available that would permit a more specific assignment elsewhere. Similarly, in interpreting statistics based on the ICD, some conditions assigned to an apparently specified category will not have been so specified on the record that was coded. When comparing trends over time and interpreting statistics, it is important to be aware that assumptions may change from one revision of the ICD to another. For example, before the Eighth Revision, an unqualified aortic aneurysm was assumed to be due to syphilis.

NOT ELSEWHERE CLASSIFIED

The words "not elsewhere classified", when used in a three-character category title, serve as a warning that certain specified variants of the listed conditions may appear in other parts of the classification. For example:

J16 Pneumonia due to other infectious organisms, not elsewhere classified.

This category includes J16.0 Chlamydial pneumonia and J16.8 Pneumonia due to other specified infectious organisms. Many other categories are provided in chapter 10 (for example, J10 – J15) and other chapters (for example, P23. - Congenital pneumonia) for pneumonias due to specified infectious organisms. J18 Pneumonia, organism unspecified, accommodates pneumonias for which the infectious agent is not stated.

"AND" IN TITLES

"And" stands for "and/or". For example, in the rubric A18.0 Tuberculosis of bones and joints, are classified cases of "tuberculosis of bones", "tuberculosis of joints" and "tuberculosis of bones and joints".

POINT DASH   .-

In some cases, the fourth character of a subcategory code is replaced by a dash, e.g.   G03 Meningitis due to other unspecified causes,

> Excludes: Meningoencephalitis (G04.-)

This indicates to the coder that a fourth character exists and should be sought in the appropriate category. This convention is used in both the tabular list and the alphabetical index.

CONVENTIONS USED IN THE ALPHABETICAL INDEX (VOL 3)

PARENTHESES

Parentheses are used in the Index in the same way as in volume 1, i.e. to enclose modifiers.

"NEC"

NEC (Not Elsewhere Classified) indicates that specified variants of the listed condition are classified elsewhere, and that, where appropriate, a more precise term should be looked for in the index.

CROSS – REFERENCES

Cross – references are used to avoid unnecessary duplication of terms in the index. The word "see" requires the coder to refer to the other term: "see also" directs the coder to refer elsewhere in the index if the statement being coded contains other information that is not found indented under the term to which "see also" is attached.

SPECIAL SIGNS USED IN THE INDEX

The following special signs are attached to certain code numbers or index terms:

†/* - Used to designate the etiology code and the manifestation code respectively, for terms subjects to dual classification.

#◊ - Attached to certain terms in the list of sites under "Neoplasm" to refer the coder to Notes 2 and 3, respectively, at the start of that list.

CHAPTER EIGHT

INTRODUCTION TO MEDICAL BILLING

By now you've got a good idea about the practice of medical coding. But we still don't know much about what those codes are used for.

While it's true that we can use diagnosis and procedure codes to track the spread of disease or the effectiveness of a particular procedure, their main use in the United States is in the reimbursement process. In other words, codes help us bill accurately and efficiently.

WHY WE BILL

Going to the doctor may seem like a one-to-one interaction, but in reality it's part of a large, complex system of information and payment. While the insured patient may only have direct interaction with one person or healthcare provider, that check-up is actually part of a three-party system.

The first party is the patient. The second party is the healthcare provider. The term 'provider' includes hospital, physicians, physical therapists, emergency rooms, outpatient facilities, and any other place where medical services are performed. The third and final party is the insurance company, or payer.

It's the medical biller's job to negotiate and arrange for payment between these three parties. Specifically, the biller ensures that the healthcare provider is compensated for their services by billing both patients and payers. We bill because healthcare providers need to be compensated for the services they perform.

In order to do this, the biller collects all of the information (found in a "super-bill") about the patient and the patient's procedure, and compiles that into a bill for the insurance company. This bill is called a claim, and it contains a patient's demographic information, medical history, and insurance coverage, in addition to a report on what procedures were performed and why.

MORE ABOUT INSURANCE

Let's take a quick step back to talk briefly about the insurance process. Health insurance is insurance against medical expenses. Put simply, people with health insurance, sometimes called 'the insured' or 'subscribers,' pay a certain amount in order to have a degree of protection against medical costs.

Health insurance comes in a number of forms, including:

Indemnity: Or pay-for-service insurance, in which the patient may choose any provider they like. This insurance is typically costlier, but grants the insured person more flexibility. As healthcare prices rise, indemnity insurance is becoming less and less popular.

Managed care organizations (MCO): This is a blanket term that includes organizations like Healthcare Maintenance Organizations (HMOs) and Preferred Provider Organizations (PPOs). Patients have fewer options as to which providers they can see, but their premiums and deductibles are fixed and are generally lower. Essentially, managed care insurance restricts patient's options but also lowers the cost of having health insurance. This is the most popular form of health insurance in the United States today.

Consumer-driven health plans (CDHP): This plan is similar to a PPO, but it also features a savings account, which subscribers pay into regularly and which is used to pay medical bills before the deductible has been met. CDHPs have high deductibles and low premiums, and are an increasingly popular option.

With each of these types of insurance, there are procedures and services that are covered, and some that are not. It's the medical biller's job to interpret a patient's insurance plan (or plans) and use this information to create an accurate claim.

MORE ABOUT CLAIMS

The creation of the claim is where medical billing most directly overlaps with medical coding. Medical billers take the procedure and diagnosis codes used by medical coders and use them to create claims. If you'll remember from Section 2, it's the coder's job to translate the medical report accurately into numeric and alphanumeric codes.

Procedure codes, whether Current Procedure Terminology (CPT) or Healthcare Common Procedure Coding System (HCPCS), tell the payer what service the healthcare provider performed. Diagnosis codes, documented using ICD codes, demonstrate medical necessity. In other words, procedure codes tell the what of a patient's visit, and the diagnosis codes tell the why.

The biller adds information about the patient and the patient's visit, along with the cost of the procedure or procedures performed, to the claim. So the claim now has a what, a why, a who, a when, and a how much.

CROSSWALKING

At this point, the biller also checks to make sure a claim is compliant. That is, the claim is factually and formally correct. This is a complicated process, as the biller must know what.

The claim allows the payer to fully evaluate the procedure and decide how much they will reimburse the provider. If the claim is approved, it's sent back to the biller with the amount the payer is going to pay. The biller then takes the amount, called the balance, and sends it on to the patient.

We'll take a closer look at The Medical Billing Process and the claims process specifically in More About Insurance and the Insurance Claims Process in this section.

DAY-TO-DAY ACTIVITIES

Now that you've got a little more information about the overall process, here's a quick look at the day-to-day activities of a professional medical biller.

WORKING WITH PATIENTS

When a patient receives medical services from a healthcare provider, they're typically presented with a bill at the end of their services. The biller creates this bill by looking at the balance (if any) the patient has, adding the cost of the procedure or service to that balance, deducting the amount covered by insurance, and factoring in a patient's copay or deductible.

Billers also work daily with a patient's medical records. Like coders, billers abstract a large amount of information from medical documents. Where coders use medical reports to accurately translate medical services into code, billers abstract information from patients' medical records and insurance plans to create accurate medical bills.

WORKING WITH COMPUTERS

Computer programs are invaluable tools in the world of medical billing. Today, almost every doctor's office in the country uses some form of practice management software. This software keeps track of patients, helps schedule visits, stores important medical information and generally helps the practice run smoothly. Medical billers use practice management software to generate reports on the status of claims, record payments from payers and patients, create statements for patients, and much more.

CREATING CLAIMS

The majority of a medical biller's day is spent creating and processing medical claims. Billers need to be familiar with what type of claim an insurance payer accepts, and adjust their claim creation accordingly. Billers may also work frequently with insurance clearinghouses to streamline the claims process. Billers also have to check that each claim is compliant. Ideally, every claim a biller sends out will be "clean." A clean claim contains no errors, and will be processed speedily by the payer, ensuring that the healthcare provider gets reimbursed quickly and efficiently.

NOTIFICATION AND COMMUNICATION

A biller is constantly in communication with insurance payers, clearinghouses, providers, and patient. Since the biller acts as the way point for the reimbursement process, they frequently have to clarify and follow-up with all parties of the healthcare process.

Billers also explain and notify patients of their bill. Billers are in charge of issuing Explanations of Benefits (EOBs) to patients, which list which procedures are covered by the payer and why. Billers must also follow up with patients about paying the balance on their medical bills. This may lead to collections.

COLLECTIONS

In the case of a patient with delinquent bills, a medical billing specialist may have to arrange for collections on that debt. This is not necessarily a "day-to-day" activity, as one would hope that a provider's patients were not ignoring their medical bills on a daily basis, but it is something to be aware of.

MEDICAL BILLING VOCABULARY

The profession of medical billing has its own specific vocabulary. Learn about some of the key terms and concepts in the medical billing field. We'll expand on a number of these topics in later courses, so just try and take in the general idea behind these terms.

Allowed Amount: The amount an insurance company will pay to reimburse a healthcare service or procedure. The patient will typically pay the balance if there is any remainder.

Ancillary Services: A service administered in a hospital or other in-patient facility beyond simple room and board. This includes physical therapy, consultations, diagnostic tests and other important medical procedures.

Appeal: The process by which a patient or provider attempts to persuade an insurance payer to pay for more (or, in certain cases, pay for any) of a medical claim. The appeal on a claim only occurs after a claim has either been denied or rejected (See "Rejected Claim" and "Denied Claim").

Applied to Deductible (ATD): The amount of money a patient owes a healthcare provider that goes to paying their annual deductible (See "Deductible"). A patient's deductible varies, and depends on that patient's insurance policy.

Assignment of Benefits (AOB): Insurance payments paid directly to the healthcare provider for medical services administered to the patient. The assignment of benefits occurs after a claim has been successfully process.

Authorization: In certain cases, a patient's insurance plan requires them to get permission from the payer before receiving a certain medical service. If a patient ignores this authorization, the claim for that procedure may be denied and the patient will be saddled with the entire bill.

Beneficiary: The person who receives benefits or insurance coverage. Beneficiaries are not always the ones paying for the plan, as in the case of children on their parent's healthcare plans.

Capitation: An arrangement between a healthcare provider and an insurance payer that pays the provider a fixed sum for every patient they take on. Capitated arrangements typically occur within HMOs (See "Health Maintenance Organization (HMO)"). HMOs enlist patients to service providers, who are paid a certain amount based on the patient's health risks, age, history, race, etc.

Clean Claim: A claim received by an insurance payer that is free from errors and processed is a timely manner. Clean claims are a huge boon to providers, as they reduce turnaround time for the reimbursement process and lower the need for time-consuming appeals processes. Many providers send their claims to third parties, like clearinghouses (See "Clearinghouse"), that specialize in creating clean claims.

Clearinghouse: A third-party organization in the billing process, and separate from the healthcare provider and the insurance payer. Clearinghouses review, edit, and format claims before sending them to insurance payers. This process is sometimes called "scrubbing."

Co-insurance: A type of insurance arrangement between the payer and the patient that divides the payment for medical services by percentage. While this is some-times used synonymously with a co-pay (See: "Co-pay"), the arrangements are different: while a co-pay is a fixed amount the patient owes, in a co-insurance, the patient owes a fixed percentage of the bill. These percentages are always listed with the payer's percentage first (eg a 70-30 co-insurance).

Coordination of Benefits: When a patient is covered by more than one insurance companies, those companies arrange themselves into a hierarchy. One payer becomes the primary carrier, and the remaining companies will assume the roles of secondary or tertiary carriers. These secondary or tertiary carriers may cover what costs are left over after the primary carrier reimburses the healthcare provider for the services rendered.

Co-pay: The amount a patient must pay to a provider before they receive any medical service. Co-pays are distinct from deductibles (See "Deductible") and are slightly different from co-insurances. The co-pay for a patient may change depending on the patient's plan and the medical service to be administered.

Crossover Claim: When a claim is sent from a primary insurance carrier to a secondary carrier, or vice versa, this is called a crossover claim.

Deductible: The amount a patient must pay before an insurance company extends their coverage. This number, which you can think of as a threshold of payment, varies depending on a patient's insurance plan. A patient with a $200 deductible, for example, would have to pay the first $200 of a $500 procedure, after which his insurance company would cover the rest. Note that this is distinct from a co-pay (See "Co-pay"), and that patients may often have to pay both their deductible and their co-pay before receiving a service.

Electronic Claim: A claim sent electronically using a provider's billing software. Electronic billing is a rapidly expanding field, but you should note claims must still adhere to billing regulations laid out by the federal government.

Explanation of Benefits (EOB): A document attached to a processed claim that explains to the provider and patient which services an insurance company will cover. EOBs may also explain what is wrong when a claim is denied.

Electronic Remittance Advice (ERA): A digital version of the EOB, this document describes how much of a claim the insurance company will pay and, in the case of a denied claim, explains why the claim was returned.

Financial Responsibility: Financial responsibility describes which party—insurance payer or patient—owes money to the healthcare provider. Financial responsibility is outline in the patient's healthcare insurance agreement.

Fiscal Intermediary (FI): A Medicare representative who processes Medicare claims.

Guarantor: An individual paying for the insurance plan who is not also the patient. Parents are the most common examples of guarantors. You may also see guarantors referred to as "responsible parties."

Health Insurance Portability and Accountability Act (HIPAA): A law passed in 1996 (USA applicable) that has lasting effects on the healthcare industry today. Title I of the act protects workers' health insurance when they change or lose jobs. Title II of the Act established standards and best practices in electronic health care.

Health Maintenance Organization (HMO): A network of healthcare providers that offer coverage to patients for medical services exclusively within that network.

Indemnity: Also known as fee-for-service insurance, this type of insurance allows patients to receive care from any healthcare provider in exchange for higher fees and deductibles. Unlike an HMO, this plan allows for greater flexibility on the patient's part, but it does cost significantly more.

Independent Practice Association (IPA): A professional organization of physicians or healthcare providers who have a contract with an HMO. HMOs contract IPAs to provide services to patients within the HMO's network, but their individual practices do not have to be part of the HMO network.

Managed Care Plan: A type of insurance plan wherein patients are only eligible to receive health care within the insurance company's network. HMOs and IPAS (See "Health Maintenance Organization (HMO)" and "Independent Practice Association (IPA)") are examples of the managed care system.

Non-covered Charge (N/C): These are procedures or services on a claim that are not covered by a person's insurance plan.

Patient Responsibility: This is the amount a patient owes the healthcare provider after an insurance payer reimburses their portion of the claim. This may also be called the balance of the bill.

Primary Care Physician (PCP): The physician that provides basic medical services for the patient, like general evaluation, low-level injuries and non-serious illnesses. The PCP may also recommend other healthcare providers to the patient. In HMOs, many PCPs act as "gatekeepers," assessing patients in the network and then sending them to the appropriate specialist in the HMO network.

Point of Service (POS) Plan: In this insurance plan, a patient in an HMO network can go to a physician outside of their network if they are referred there and pay a higher deductible. Think of this as a cross between an HMO and basic indemnity insurance (See "Health Maintenance Organization" and "Indemnity").

Preferred Provider Organization (PPO): A plan similar to an HMO, except that the insurance company, rather than the HMO itself, decides who is in the acceptable provider network. This is a common, subscription-based type of managed care.

Premium: This is the amount a patient regularly pays to an insurance company in order to receive coverage. Premiums are typically paid on a monthly or yearly basis.

Provider: Any healthcare facility that administers healthcare to an individual. Physicians, specialists, clinics, hospitals, general practitioners, and outpatient facilities are all considered providers.

Specialist: A provider, either an individual or an office, that focuses on one type of healthcare. Oncologists, physical therapists, and ophthalmologists are all ex-amples of specialists. In many cases, a patient needs to be referred to a specialist by a primary care physician (See "Primary Care Physician (PCP)") before seeing a specialist for the first time, especially if that patient is a member of a managed care network (See "Managed Care").

Subscriber: The person who is covered under a group policy. Members of managed care networks are subscribers to that network (See "Managed Care").

Super bill: Used by healthcare providers, this is an itemized account of the provider's encounter with a patient. The super bill is the main source of data for creating the medical claim, and may include demographic information, insurance information, diagnoses, and procedures performed.

Supplemental Insurance: A secondary or auxiliary insurance policy that covers a patient's healthcare cost after they receive coverage from their primary coverage. Supplemental insurance may also be called secondary or, in the case of a patient having more than two policies, tertiary coverage. These supplemental insurance plans are often put in place to help patients cover high deductibles or co-pays.

Triple Option Plan (TOP): Sometimes called a "cafeteria plan," this plan provides individuals who sign up the option of choosing between an HMO, PPO, or POS coverage (See "Health Maintenance Organization (HMO)," "Preferred Provider Organization (PPO)," and "Point of Service (POS) Plan").

Untimely Submission: Claims must be filed within a certain time frame. Think of this as an expiration date. If a claim is not sent to an insurance company within the designated time frame, this claim is labeled an untimely submission and will be denied.

Utilization Limit: Medicare places a yearly limit on certain medical services. If a patient passes this threshold, known as the utilization limit, they may be ineligible for Medicare coverage for that procedure.

Worker's Compensation: When a company pays for the health insurance of an employee who becomes injured or ill while performing their job's routine duties. Most states require companies to provide worker's compensation to their employees.

THE MEDICAL BILLING PROCESS

Medical billing might seem large and complicated, but it's actually a process that's comprised of eight simple steps.

These steps include: Registration, establishment of financial responsibility for the visit, patient check-in and check-out, checking for coding and billing compliance, preparing and transmitting claims, monitoring payer adjudication, generating patient statements or bills, and assigning patient payments and arranging collections. Bear in mind that there is a difference between "front-of-house" and "back-of-house" duties when it comes to medical billing.

REGISTER PATIENTS

When a patient calls to set up an appointment with a healthcare provider, they effectively preregister for their doctor's visit. If the patient has seen the provider before, their information is on file with the provider, and the patient need only explain the reason for their visit. If the patient is new, that person must provide personal and insurance information to the provider to ensure that that they are eligible to receive services from the provider.

CONFIRM FINANCIAL RESPONSIBILITY

Financial responsibility describes who owes what for a particular doctor's visit. Once the biller has the pertinent info from the patient, that biller can then determine which services are covered under the patient's insurance plan.

Insurance coverage differs dramatically between companies, individuals, and plans, so the biller must check each patient's coverage in order to assign the bill correctly. Certain insurance plans do not cover certain services or prescription medications. If the patient's insurance does not cover the procedure or service to be rendered, the biller must make the patient aware that they will cover the entirety of the bill.

PATIENT CHECK-IN & CHECK-OUT

Patient check-in and check-out are relatively straight-forward front-of-house procedures. When the patient arrives, they will be asked to complete some forms (if it is their first time visiting the provider), or confirm the information the doctor has on file (if it's not the first time the patient has seen the provider). The patient will also be required to provide some sort of official identification, like a driver's license or passport, in addition to a valid insurance card.

The provider's office will also collect co-payments during patient check-in or check-out. Co-payments are always collected at the point of service, but it's up to the provider to determine whether the patient pays the copay before or immediately after their visit.

Once the patient checks out, the medical report from that patient's visit is sent to the medical coder, who abstracts and translates the information in the report into accurate, useable medical code. This report, which also includes demo-graphic information on the patient and information about the patient's medical history, is called the "**superbill**."

SUPERBILL

The **superbill** contains all of the necessary information about medical service provided. This includes the name of the provider, the name of the physician, the name of the patient, the procedures performed, the codes for the diagnosis and procedure, and other pertinent medical information. This information is vital in the creation of the claim. Once complete, the superbill is then transferred, typically through a software program, to the medical biller.

PREPARE CLAIMS /CHECK COMPLIANCE

The medical biller takes the superbill from the medical coder and puts it either into a paper claim form, or into the proper practice management or billing soft-ware. Billers will also include the cost of the procedures in the claim. They won't send the full cost to the payer, but rather the amount they expect the payer to pay, as laid out in the payer's contract with the patient and the provider.

Once the biller has created the medical claim, he or she is responsible for ensuring that the claim meets the standards of compliance, both for coding and format.

The accuracy of the coding process is generally left up to the coder, but the biller does review the codes to ensure that the procedures coded are billable. Whether a procedure is billable depends on the patient's insurance plan and the regulations laid out by the payer.

While claims may vary in format, they typically have the same basic information.Each claim contains the patient information (their demographic info and medical history) and the procedures performed (in CPT or HCPCS codes). Each of these procedures is paired with a diagnosis code (an ICD code) that demonstrates the medical necessity. The price for these procedures is listed as well. Claims also have information about the provider, listed via a National Provider Index (NPI) number. Some claims will also include a Place of Service code, which details what type of facility the medical services were performed in.

Billers must also ensure that the bill meets the standards of billing compliance. Billers typically must follow guidelines laid out by the Health Insurance Portability and Accountability Act (HIPAA) and the Office of the Inspector General (OIG). We'll discuss HIPAA and its effect on medical billing in Course 3-8 and 3-9. OIG compliance standards are relatively straightforward, but lengthy, and for reasons of space and efficiency, we won't cover them in any great depth here.

TRANSMIT CLAIMS

Since the Health Insurance Portability and Accountability Act of 1996 (HIPAA), in the United States of America (USA)   all health entities covered by HIPAA have been required to submit their claims electronically, except in certain circumstances. Most providers, clearinghouses, and payers are covered by HIPAA.

Note that HIPAA does not require physicians to conduct all transactions electronically. Only those standard transactions listed under HIPAA guidelines must be completed electronically. Claims are one such standard transaction.

Billers may still use manual claims, but this practice has significant drawbacks. Manual claims have a high rate of errors, low levels of efficiency, and take a long time to get from providers to payers. Billing electronically saves time, effort, and money, and significantly reduces human or administrative error in the billing process.

In the case of high-volume third-party payers, like Medicare or Medicaid, billers can submit the claim directly to the payer. If, however, a biller is not submitting a claim directly to these large payers, they will most likely go through a clearinghouse.

A clearinghouse is a third-party organization or company that receives and re-formats claims from billers and then transmits them to payers. Some payers require claims to be submitted in very specific forms. Clearinghouse eases the burden of medical billers by taking the information necessary to create a claim and then placing it in the appropriate form. Think of it this way: A practice may send out ten claims to ten different insurance payers, each with their own set of guidelines for claim submission. Instead of having to format each claim specifically, a biller can simply send the relevant information to a clearinghouse, which will then handle the burden of reformatting those ten different claims.

MONITOR ADJUDICATION

Once a claim reaches a payer, it undergoes a process called adjudication. In adjudication, a payer evaluates a medical claim and decides whether the claim is valid/compliant and, if so, how much of the claim the payer will reimburse the provider for. It's at this stage that a claim may be accepted, denied, or rejected.

A quick word about these terms. An accepted claim is, obviously, one that has been found valid by the payer. Accepted does not necessarily mean that the payer will pay the entirety of the bill. Rather, they will process the claim within the rules of the arrangement they have with their subscriber (the patient).

REJECTED CLAIM

A rejected claim is one that the payer has found some error with. If a claim is missing important patient information, or if there is a miscoded procedure or diagnosis, the claim will be rejected, and will be returned to the provider/biller. In the case of rejected claims, the biller may correct the claim and resubmit it.

DENIED CLAIM

A denied claim is one that the payer refuses to process payment for the medical services rendered. This may occur when a provider bills for a procedure that is not included in a patient's insurance coverage. This might include a procedure for a preexisting condition (if the insurance plan does not cover such a procedure).

Once the payer adjudication is complete, the payer will send a report to the provider/biller, detailing what and how much of the claim they are willing to pay and why. This report will list the procedures the payer will cover and the amount payer has assigned for each procedure. This often differs from the fees listed in the initial claim. The payer usually has a contract with the provider that stipulates the fees and reimbursement rates for a number of procedures. The report will also provide explanations as to why certain procedures will not be covered by the payer.

(If the patient has secondary insurance, the biller takes the amount left over after the primary insurance returns the approved claim and sends it to the patient's secondary insurance).

The biller reviews this report in order to make sure all procedures listed on the initial claim are accounted for in the report. They will also check to make sure the codes listed on the payer's report match those of the initial claim. Finally, the biller will check to make sure the fees in the report are accurate with regard to the contract between the payer and the provider.

If there are any discrepancies, the biller/provider will enter into an appeal pro-cess with the payer. This process is complicated and depends on rules that are specific to payers and to the states in which a provider is located. Effectively, a claims appeal is the process by which a provider attempts to secure the proper reimbursement for their services. This can be a long and arduous process, which is why it's imperative that billers create accurate, "clean" claims on the first go.

GENERATE PATIENT STATEMENTS

Once the biller has received the report from the payer, it's time to make the statement for the patient. The statement is the bill for the procedure or procedures the patient received from the provider. Once the payer has agreed to pay the provider for a portion of the services on the claim, the remaining amount is passed to the patient.

In certain cases, a biller may include an Explanation of Benefits (EOB) with the statement. As we learned in the previous course, an EOB describes what benefits, and therefore what kind of coverage, a patient receives under their plan. EOBs can be useful in explaining to patients why certain procedures were covered while others were not.

FOLLOW UP ON PATIENT PAYMENTS AND HANDLE COLLECTIONS

The final phase of the billing process is ensuring those bills get, well, paid. Billers are in charge of mailing out timely, accurate medical bills, and then following up with patients whose bills are delinquent. Once a bill is paid, that information is stored with the patient's file.

If the patient is delinquent in their payment, or if they do not pay the full amount, it is the responsibility of the biller to ensure that the provider is properly reimbursed for their services. This may involve contacting the patient directly, sending follow-up bills, or, in worst-case scenarios, enlisting a collection agency.

Each provider has it's own set of guidelines and timelines when it comes to bill payment, notifications, and collections, so you'll have to refer to the provider's billing standards before engaging in these activities.

MORE ABOUT INSURANCE AND THE INSURANCE CLAIMS PROCESS

Health care is, as many have noted, one of the largest and fastest-growing sectors of the American economy. Americans spend almost $8,000 annually per capita on healthcare, and a significant portion of that sum is spend on health insurance.

HOW HEALTH INSURANCE WORKS

Health insurance is insurance against medical expenses. Essentially, health insurance subscribers enter into an arrangement with a health insurance company in order to reduce the impact of the cost of medical expenses. There are many different types of insurance coverage plans, and even more ways of paying for them.

Most plans share a few basic similarities. Most insurance plans require subsc-ribers to pay premiums, which are essentially subscription fees. These may be assigned monthly or annually.

EXAMPLE

Many plans also have deductibles, which are monetary limits after which the health insurance company assumes the cost of the medical procedure or service. For instance, if a person has an insurance plan with a $100 deductible, he will pay up to $100 for a medical procedure, and his insurance company will pay for the remaining amount (provided that procedure is valid and within their insurance arrangement).

Subscribers may also have a copay or coinsurance arrangement with their insurance company. A copay is a relatively small, fixed sum that must be paid before any medical service is rendered. The co-pay does not count against the deductible. So, if that same patient has a ₦100 deductible and a ₦25 co-pay for a particular procedure, the patient will have to pay the ₦25 co-pay first, and then the ₦100 deductible, after which point the insurance company will pay the rest.

A co-insurance is a type of arrangement with the insurance company that divides the responsibility for payment by percentage. Co-insurances are listed with the payer (insurance company)'s portion listed first, and then the subscriber's. For instance, if a subscriber receives a ₦3000 medical procedure,

and has a 80-20 co-insurance agreement with his or her insurance company, the subscriber would owe 20% of the bill (₦600). The insurance company would pay the rest.

Co-insurances also come with deductibles. Looking at the example above, let's say the subscriber has a ₦1000 deductible in addition to his 80-20 co-insurance plan. If he received a ₦3000 procedure, he'd have to first pay his deductible, and then 20% of the remaining figure.

Here's a breakdown:

₦3000 (total cost of procedure)

–₦1000 (deductible)

₦200 (remaining amount left to pay)

20% of ₦2000 (subscriber's co-insurance rate) = $400

So the subscribers total amount would be ₦1400. The insurance company would pay the remaining 80% of the ₦2000, which would come to ₦1600. Now that we've got an idea of how some of the basic aspects of health insurance work, let's take a look at the different types of health insurance.

INDEMNITY

We covered indemnity insurance briefly in Course 3-1, but let's return to it now for the sake or review. Indemnity is the most basic and straightforward kind of insurance, in that you pay a premium to an insurance company to insulate you from medical expenses. You'll likely have a deductible and, depending on your insurance plan, a co-pay or co-insurance. Subscribers to indemnity plans have no restrictions on which providers they can see, but indemnity plans are typically much more expensive than managed care options, which we'll review now.

MANAGED CARE

We touched briefly on managed care and managed care organizations in our introduction to this section. Let's revisit these now. Managed Care Organizations (MCOs) are groups, organizations, or other bodies that seek to reduce the cost of healthcare and increase the efficacy or health services through a number of means.

Managed care organizations, for instance, may confine the providers the subscriber may see to a specific network of doctors and facilities. In general, MCOs have fixed costs that are lower than most indemnity plans, but restrict the options a patient has for where to get treatment.

MANAGED CARE ORGANIZATIONS

There are three main types of MCO, which we'll discuss below. Bear in mind that these are simplified descriptions of these managed care organizations.

Health Management Organization (HMO) – At one time, HMOs were the most popular MCO option. HMOs operate by providing subscribers with a low premium and a strict network of providers a subscriber can see. If a subscriber sees a provider outside fo this network, they may have to cover all of the expenses from that service out of pockets. HMOs are often among the cheapest MCOs, but are also the least flexible. HMOs also often make use of primary care physicians (PCPs), who may act as "gatekeepers." Subscribers often need to be referred to specialists by PCPs.

Preferred Provider Organization (PPO) – PPOs recently over took HMOs as the most common MCO. Unlike an HMO, subscribers to a PPO may see any doctor, physician or other provider, but they pay less if they see a provider within the PPO's network (hence "preferred"). PPOs generally have higher premiums, but allow for more flexibility for subscribers.

Point of Service (POS) – A slight variation on the HMO model, subscribers to a POS plan fulfil most of the medical needs in-network, but are allowed to go out-of-network if they pay a higher fee. Many POS plans are tiered, so that a subscriber pays more if they see a specialist out-of-network,

but less if they are referred to that specialist by an in-network PCP.

Consumer-Driven Health Plan (CDHP) – A relatively recent development in the world of MCOs, CDHPs enable subscribers to receive PPO-like benefits only after they've paid a certain deductible. This deductible is usually quite high, but comes with low premiums and a "savings account" that works like a retirement fund. Subscribers may put money into the account to help pay for out-of-pocket expenses. Why do we need to know about all these different types of insurance coverage? Because each of these affects the way we create claims.

BILLING EXAMPLE

Let's say we're billing for a procedure that cost ₦1500. The patient who received the procedure has a CDHP with a deductible of ₦1000. In order to create an accurate claim, we'd look at the patient's coverage plan, and assign the ₦1000 deductible to the patient, and then pass the ₦500 on to the payer.

Likewise, if we're looking at a patient with coverage under an HMO, but that patient sees a provider out-of-network, we need to know that we can't send a claim to that HMO, but must instead bill the patient directly. (Recall that HMO subscribers cannot receive insurance coverage if they see providers out of their network).

Knowing the ins and outs of insurance plans—what type of coverage they provide, how much to deduct and send to the payer—is an integral part of the billing process.

POTENTIAL BILLING PROBLEMS AND RETURNED CLAIMS

The goal of the medical biller is to ensure that the provider is properly reimbursed for their services. In the pursuit of this goal, errors, both human and electronic, are unfortunately unavoidable. Since the process of medical billing involves two incredibly important elements (namely, health and money), it's important to reduce as many of these errors as possible. In these brief course, we'll introduce you to some common errors in the medical billing practice.

Before we jump into that discussion, however, let's review the difference between a rejected and denied claim.

DENIED AND REJECTED CLAIMS

As you'll recall from previous Courses, a rejected claim is not the same as a denied one. A rejected claim is one that contains one or many errors found before the claim is processed. These errors prevent the insurance company from paying the bill as it is composed, and the rejected claim is returned to the biller in order to be corrected. A rejected claim may be the result of a clerical error, or it may come down to mismatched procedure and ICD codes. A rejected claim will be returned to the biller with an explanation of the error. These claims are then corrected and resubmitted.

Clearinghouses employ a process colloquially called "scrubbing" in order to avoid rejected claims. The end goal, for billers and clearinghouses, is a "clean" claim.

Denied claims, on the other hand, are claims that the payer has processed and deemed unpayable. These claims may violate the terms of the payer-patient contract, or they may just contain some sort of vital error that was only caught after processing. Payers will include an explanation for why a claim is denied when they send the denied claim back to the biller. Many times, these claims can be appealed and sent back to the payer for processing, but this process can be time-consuming and, therefore, costly. For that reason, it's important to try and get as many claims "clean" on the first go, and not waste any time billing for procedures that are incompatible with a patient's coverage.

SIMPLE ERRORS

Now that we've reviewed denied and rejected claims, let's look at some of the basic errors that can get a claim returned to the biller.

- ***Incorrect patient information***

    Sex, name, DOB, insurance ID number, etc.

- ***Incorrect provider information***

    Address, name, contact information, etc.

- ***Incorrect Insurance provider information***

    Wrong policy number, address, etc

- ***Incorrect codes***

    entering confusing ICD, CPT, or HPCS codes; entering confusing Place of Service codes;   attaching   conflicting   or   confusing modifiers to HCPCS or CPT codes; entering too few or   too   many digits to an ICD, CPT, or HCPCS codes

- ***Mismatched medical codes***

    Entering confusing ICD codes with CPT codes, or vice versa, etc

- ***Leaving out codes altogether for procedures or diagnoses***

- **Duplicate Billing**

This occurs when someone at the provider's office submits a claim for a procedure without checking whether that service has been paid for/reported. Duplicate billing can create a huge headache for billers and payers alike, because it may appear that a patient received two identical x-rays on one day, which would effectively double the amount sent to the payer.

Like medical coding, we're always striving for the highest level of accuracy in our codes, and we're also required to provide as complete a picture as possible of the medical procedure(s). If you can cut down on these simple errors in your medical billing, you'll have a much higher number of clean claims.

MORE BILLING ERRORS

The above are some of the most frequent errors a medical biller comes across. These errors directly affect the status of a claim, which makes them very important to watch out for.

But there are other errors to watch out for as you go through your day as a medical biller. Some of these are, regrettably, out of the biller's hands, but they're important to watch out for nonetheless.

- ## Undercoding

Undercoding occurs when a provider intentionally leaves out a procedure code from a superbill, or codes for a less serious or extensive procedure than the patient received. Undercoding may be done to avoid audits for certain procedures, or to try and save money for the patient. This process is illegal, and counts as a type of fraud.

- ## Upcoding

Like undercoding, this is a fraudulent process wherein the provider intentionally misrepresents the work they performed on a patient. In upcoding, a practice enters codes for services a patient did not receive, or codes for more intensive procedures then the provider actually performed. Upcoding is typically done in an attempt to receive more money from a payer. This, like undercoding, is a fraudulent practice, and should be noted and reported immediately.

- ## Poor documentation

While not a fraudulent practice like upcoding or undercoding, poor documentation can also negatively affect the claims process. If a provider has provided incorrect, illegible, or incomplete documentation of a procedure or patient visit, it's difficult to make an accurate or complete claim. In cases of sloppy documentation, the biller should contact the provider and ask for more information.

- **No EOB on denied claim**

In certain cases, the payer may fail to attach the Explanation of Benefits (EOB) to a denied claim. In cases like this, it's difficult to note the error on a denied claim, which slows down the (already slow) appeals process.

FIXING ERRORS BEFORE THEY HAPPEN

It's always important to be proactive when you're medical billing. Here are a couple of things you can do to catch medical billing errors before they happen.

- **Stay Current**

•Billers need to stay up-to-date on billing and coding trends. Coding especially will change as new codes are introduced and older ones phased out. It's important to check on new protocols in medical coding regularly. Study new codes and be aware of how they affect billing.

- ## Be Diligent

You should always double check your work when you're creating a claim. Simple clerical   errors like missing digits or misspelled names can be the difference between an approved and a rejected claim, so go over each claim you create before you send it   off.

- ## Communicate

Part of reducing medical billing errors comes down to coordinating effectively within   the provider's office. Make sure you communicate regularly and effectively with other   personnel in   the provider's office, including the physician, and don't be afraid to ask questions about possible errors on the claim.

- ## Follow Through

After you send a claim in to a payer, you can follow up with a representative working on that   claim. They may be able to alert you to any errors they've already caught, in which case you can   begin work on making a new, error-free claim. (Wait until they send it back to you, of course!)

CHAPTER NINE

SECONDARY RECORDS AND DATABASES
(INDEXES AND REGISTRIES)

INTRODUCTION

Health record is a rich source of data about an individual patient, and it fulfills the primary use of patient care and reimbursement for individual patient encounters. However, it is not easy to see the trends in a population of patients by looking at individual records. For this purpose, data must be extracted from individual records and entered into specialized databases that support analysis across individual records. These data may be used in a **facility – specific or population – based** registry for **research** and **improvement in patient care**. In addition, they may be reported to the state and become part of the state and federal – level databases that are **used to set health policy and improve health care**.

The health information management (HIM) professionals can play a variety of roles in managing secondary records and databases. He or she plays a key role in helping to set up databases. This task includes determining the content of the database or registry and ensuring compliance with the laws, regulations, and accrediting standards that affect the content and use of the registry and database. All data elements included in the database or registry must be defined in a data dictionary. In this role, the HIM professional may oversee the completeness and accuracy of the data abstracted for inclusion in the database or registry.

This chapter explains the difference between primary and secondary data sources and their users. It also offers an in-depth look at the types of secondary data bases, including indexes, registries, and their functions.

PRIMARY VERSUS SECONDARY DATA SOURCES AND DATABASES

The health record is considered a primary data source because it contains information about a patient that has been documented by the professionals who provided care or services to that patient. Data taken from the primary health records and entered into registries and databases are considered a secondary data source.

Data also are considered as either patient – specific /identifiable data or aggregate data. The health record consists entirely of patient – identifiable data. In other words, every fact recorded in the record relates to a particular patient identified by name. Secondary data also may be patient identifiable. In some instances, data are entered into a data base along with information such as the patient's name, maintained in an identifiable form. Registries are an example of patient-identifiable data on groups of patients.

More often, however, secondary data are considered aggregate data. Aggregate data include data on groups of people or patients without identifying any particular patient individually. Examples of aggregate data are statistics on the average length of stay (ALOS) for patient discharged within a particular diagnosis-related group (DRG).

PURPOSES AND USERS OF SECONDARY DATA SOURCES

Secondary data sources consists of facility-specific indexes; registries, either facility or population based; and other health care databases. Healthcare organizations maintain those indexes, registries, and databases that are relevant to their specific operations.

Secondary data sources provide information that is not easily available by looking at individual health records. For example, if a researcher wanted to find the records of 30 patients who had the principal diagnosis of myocardial infarction, he or she would have to look at numerous individual records to locate the number needed. This would be time-consuming and laborious project. With a diagnosis index, the task would involve simply looking at the list of diagnosis in numerical order and selecting those with the appropriate ICD-10 diagnosis code for inclusion in the study.

Data extracted from health records and entered into disease oriented database can, for example, help researchers determine the effectiveness of alternate treatment methods. They also can quickly demonstrate survival rates at different stages of disease.

Internal users of secondary data are individuals located within the health care facility. For example, internal users include medical staff and administrative and management staff. Secondary data enable these users to identify patterns and trends that are helpful in patient care, long-range planning, budgeting, and benchmarking with other facilities.

External users of patient data are individuals and institutions outside the facility. Examples of external users are state data-banks and government agencies. States have laws that cases of patients with diseases such as tuberculosis and AIDS must be reported to the state ministry of health. Moreover, the federal government collects data from the states on vital events such as births and deaths.

The secondary data provided to external users is generally aggregate data and not patient-identifiable data. Thus, these data can be used as needed without risking breaches of confidentiality.

FACILITY-SPECIFIC INDEXES

The secondary data sources that have been in existence the longest are those that have been developed within facility to meet their individual needs. These indexes enable health records to be located by diagnosis, procedure, or physician.

Prior to existence computerization in health care, these indexes were kept on cards. They now are usually computerized reports available from data included in databases routinely maintained in the healthcare facility. Most healthcare facilities maintain indexes described in the following subsections.

INDEXES

Health care facilities generally maintain the following indexes:

- Master patient index (MPI)
- Disease index
- Procedure index
- Physician index

Master Patient Index

A master patient index (MPI), sometimes called a master person index (MPI), links a patient's medical record number with common identification data elements(e.g., patient's complete name, date of birth, gender, mother's maiden name, and social security number). Because most health care facilities house patient records according to a medical record number, the MPI becomes the key to locating paper-based records in the health information department file system. Thus, the MPI is retained permanently because it serves as the "key" to finding the patient's record. It can be automated or manual.

An automated MPI resides on a computer and consists of a database of identification data about patients who have received health care services from a facility. An admission/discharge/transfer (ADT) system is used to input patient registration information (Figure 8-1), which results in the creation of an automated MPI database that allows for the storage and retrieval of the information. ADT software has the capability of generating standard reports for administrative and departmental purposes, including:
• Admission logs or register (list of patients admitted)
• Bed utilization reports (facility occupancy rates)
• Current charges reports (expected accounts receivable)
• Daily census summaries (current inpatients)
• Daily discharge logs or registers (list of patients discharged and transferred to other facilities)
• Patient profiles (based on patient demographics, diagnoses/procedures, and so on)
• Transfer reports (patients transferred to units within the facility)
• User-defined reports (based on user-defined criteria)

A manual master patient index (MPI) (Figure 8-2) requires the typing or hand posting of patient identification information on pre-printed index cards, and limited

information can be retrieved. MPI cards are housed in a vertical file (discussed in Chapter 7), with one card generated for each patient. When patients return to the

facility, the MPI card is retrieved and reviewed to verify demographic information and update the card with new admission information. File guides (Figure 8-3) help users quickly locate MPI cards and file folders by dividing the filing system into smaller subdivisions.

Advantages and disadvantages of automated and manual MPI systems include:

• Manual MPI is relatively inexpensive to purchase as compared with automated MPI, which requires initial purchase of computer equipment and software as well as software upgrades.

• Automated MPI allows for rapid retrieval of patient information, although a manual MPI allows for access when computer systems are unavailable (e.g., power outage).

• Manual MPI limits information that can be entered on each card, while automated MPI can be set up to meet the facility's specifications for data retrieval.

• Automated MPI usually allows for retrieval of patient information according to phonetic filing system (e.g., Soundex), while manual MPI cards can be lost if the patient's information was typed or recorded incorrectly.

• Manual MPI requires retrieval of information within the health information department, while automated MPI can be accessed by authorized personnel outside of the health information department.

• Automated MPI captures patient information upon admission and allows for computer interface, which is the exchange of data among multiple software products (e.g., patient billing, case abstracting).

Identification information entered into an MPI typically includes the following, sometimes referred to as demographic information:
• Patient name (last name, first name, middle initial)
• Address (street, city, state, zip code)
• Social security number (SSN)
• Date of birth (using mmddyyyy format)
• Admission/discharge (or transfer) dates (using mmddyyyy format)
• Medical record number (assigned by the facility)
• Name of facility and/or provider (when multiple facilities/providers are associated with the network)
• Type of care received (inpatient, outpatient, emergency, provider office)

MPI systems can also capture diagnosis/procedure descriptions for each date of service if the facility determines that this information should be collected.
Additional information entered may include race/ethnicity as well as the mother's maiden name and place of birth, which serves as identifying information for the purpose of verifying a patient (e.g., patients with common first and last names). When patients receive care at a number of facilities within a

health care network (e.g., privately owned health care system), the need to maintain current demographic data and synchronize that data is crucial. The MPI allows the health care network to uniquely identify a patient and allows providers to retrieve clinical information from wherever the patient has received care.

Last name	First name	Middle name	Gender	Age	Race	Patient No
Ado	**John**	**Sule**	**Male**	**40**	**A**	**123456**

Address		Birth	Day	Month	Year
8, Harvey road, Yaba Lagos		**Date**	**01**	**01**	**1977**

Mother's Maiden Name	Place of Birth	Social Security/National ID No
Smith	**Abuja**	**123-45-6789**

Admission Date	Discharge Date	Provider	Type	Discharge Status
05-05-2009	05-10-2004	Adebayo, James	IP	Home
06-10-2009		Ibrahim, Sandra	ED	Home
07-15-2010		Chimaroke, John	OP	Home

Figure: 9.1   Master Patient Index

Purpose of the Master Patient Index

The master patient index (MPI) is used administratively, for continuity of care (or continuum of care), and externally. Administratively, the MPI serves as a "customer database" for the health care organization and allows for the production of a variety of reports that can be used as business planning and marketing tools.

For continuity of care (or continuum of care), the MPI assists in determining whether a patient has been previously treated by a health care facility. This alerts the provider to request previous patient records to be sent to the inpatient unit, emergency department, outpatient clinic, and other departments. The review of

previous records allows the provider to most appropriately treat the patient. Externally, the MPI allows the facility to link patient services received outside of the

organization with community-wide ancillary services (e.g., services provided by a stand-alone laboratory). As a result, the facility avoids providing duplicate

services to patients, improves provider productivity (e.g., by making computerized test results available), and increases the possibility of detecting government

medical program fraud or abuse.

EXAMPLE OF ADMINISTRATIVE USE

The public relations (PR) department has been requested to perform an analysis of its health care facility's target patient care market. The PR department submits a

request to computing services to generate a zip code distribution report of patients treated by the facility during the past five years. This report can be analyzed to determine additional markets that the facility should target for advertising purposes.

EXAMPLE OF CONTINUITY OF CARE

A patient comes to the emergency department (ED) complaining of severe headaches. The emergency physician instructs the ED clerk to determine whether the patient
has previously been treated. The ED clerk obtains the patient's previous records, and upon review the ED physician notes that the patient underwent a head X-ray one
week ago. Thus, the physician selects a diagnostic workup and treatment modality that does not duplicate previous care provided.

EXAMPLE OF EXTERNAL USE

The board of directors of a health care facility based in an urban area researches whether building a satellite facility in the suburbs of a major city is justified. In addition to a zip code distribution report of patients treated by the facility,
the board should also review reports that contain data including patient age, diagnosis, procedures, and so on.

Disease, Procedure, and Physician Indexes

Disease, procedure, and physician indexes contain data abstracted (selected) from patient records and entered into a computerized database from which the respective index is generated. The disease index (Figure 8-7) is organized according to ICD-9-CM disease codes. The procedure index (Figure 8-8) is organized according to ICD-9-CM and/or CPT/HCPCS procedure/service codes. The physician index (Figure 8-9) is organized according to numbers assigned by the facility to physicians who treat inpatients and outpatients. Elements routinely entered into the database include the following:

• Demographic information (age, ethnicity, gender, inpatient admission/discharge or outpatient treatment date, and zip code)
• Financial information (third-party payer type and total charges)
• Medical information (attending physician, consulting

physician, surgeon, medical service classification (e.g., obstetrics), disease and/or procedure/ services code(s), date(s) of surgery, and type of anesthesia) Indexes are used to complete applications for accreditation prior to survey (e.g., The Joint Commission, documents required by licensing and regulatory agencies (e.g., CMS), medical and statistical reports (e.g., New York's Statewide Planning and Research Cooperative System or SPARCS), and facility-wide quality review studies of patient care.

All State Medical Center

Disease Index

Reporting Period   08-01-YYYY   08-01-YYYY   **Date Prepared** 08-02-YYYY

Page 1 of 5

Primary Dx	Other Diagnoses	Attending Dr	Age	Gender	Payer	Patient #
HUMAN IMMUNODEFICIENCY VIRUS [HIV] DISEASE						
042	112.0	138	24	M	BC	236248
042	136.3	024	35	M	BC	123456
042	176.0	036	42	F	BC	213654
ACUTE POLIOMYELITIS						
045	250.00	236	80	M	MC	236954
045	401.9	235	60	F	MD	562159
045	496	138	34	F	WC	236268

Figure: 9.2   Disease Index

All State Medical Center

Procedure Index

Reporting Period   0301YYYY—0301YYYY   **Date Prepared** 03-02-YYYY

Page 1 of 5

Primary Px	Other Procedures	Attending Dr	Age	Gender	Payer	Patient #
CLOSED BIOPSY OF BRAIN						
01.13		248	42	F	01	562359
CRANIOTOMY NOS						
01.24		235	56	F	03	231587
01.24		326	27	M	02	239854
01.24		236	08	F	05	562198
01.24		236	88	M	05	615789
DEBRIDEMENT OF SKULL NOS						
01.25		326	43	M	03	653218

Figure: 9.3   Procedure Index

All State Medical Center

Physician Index

Reporting Period 0101YYYY—0101YYYY　　　　　　**Date Prepared** 01-02-YYYY
　　　　　　　　　　　　　　　　　　　　　　　　　　Page 1 of 5

Attending Dr	Patient #	Age	Gender	Payer	Admission	Discharge	LOS	Dx
JAMES SMITH, M.D.								
024	123456	35	M	BC	1228YYYY	0101YYYY	4	042
024	213654	42	F	BC	1229YYYY	0101YYYY	3	042
024	236248	24	M	BC	1229YYYY	0101YYYY	3	042
JANE THOMSON, M.D.								
025	236268	34	F	WC	1229YYYY	0101YYYY	3	045
025	562159	60	F	MD	1230YYYY	0101YYYY	2	045
025	236954	80	M	MC	1231YYYY	0101YYYY	1	045

Figure: 9.4　Physician Index

REGISTRIES

Disease registries are collections of secondary data related to patients with a specific diagnosis, condition, or procedure. Registries are different from indexes in that they contain more extensive data. Index reports can usually be produced using data from the facility's existing databases. Registries often require more extensive data from the patient record. Each registry must define the cases that are to be included in it. This process is called **case definition**. In a trauma registry for example, the case definition might be all patients admitted with a diagnosis falling into ICD-10 code numbers of a certain range in the trauma diagnosis code.

After the cases to be included have been determined through the case definition process described above, the next step in data acquisition is usually **case finding**.

Case finding includes the methods used to identify the patients who have been seen and/or treated in the facility for the particular disease or condition of interest to the registry. After cases have been identified, extensive information is abstracted from the paper-based patient record into the registry database or extracted from other databases and entered into the registry database.

The sole purpose of some registries is to collect data from the patient health record and to make them available for users; other registries take further steps to enter additional information in the registry database, such as routine follow-up of patients at specified intervals.

Follow-up might include rate and duration of survival and quality of life issues overtime.

CANCER REGISTRY

Cancer registries have a long history in health care. It has long been recognized that aggregate clinical information is needed to improve the diagnosis and treatment of cancer.

Cancer registration is the continuing process of systematic collection of data on the characteristics of a cancer and of the subjects with cancer. The cancer registry is an organization for the systematic collection, storage, analysis, interpretation, and reporting of data on subjects with cancer. The registry may be **facility based** (located within a facility such as a hospital or clinic) or **population based_**(gathering information from more than one facility within a geographical defined area such as a state or region).

The data from **facility-based registries** are used to provide information for the improved understanding of cancer, including its cause and methods of diagnosis and treatment. The data collected also may provide comparisons in survival rates and quality of life for patients with different treatments and at different stages of cancer at the time of diagnosis. This implies that the data are used for the administrative purposes, and for reviewing clinical performance.

A population-based cancer registry collects data on every subject with cancer in a defined population.

Usually the population is that which is resident in a well-defined geographical region. The registry must therefore be able to distinguish between residents and non-residents, and should have sufficient information on each case to avoid multiple registrations.

In the population-based registries emphasis is on identifying trends and changes in the incidence (new cases) of cancer within the area covered by the registry.

Since population registration involves head counts of new cancer cases occurring in a defined time period and relates this to the "population at risk", the availability of dependable population census data by sex and five-year age-groups is essential.

CASE DEFINITION AND CASE FINDING IN THE CANCER REGISTRY

As defined previously, case definition is the process of deciding what cases should be entered into the registry. In a cancer registry, for example, all cancers cases except certain skin cancers might meet the definition for the cases to be included. Skin cancer such as basal cell carcinoma might be excluded because they do not metastasize and do not require follow-up necessary for other cancers included in the registry. Data on benign and borderline brain/central nervous system tumours also must be collected.

In the facility-based cancer registry, the first step is case finding. One way to find cases is through the discharge process in the HIM department. During the discharge procedure, coders and/or discharge analyst can easily earmark cases of patients with cancer for inclusion in the registry. Another case finding method is to use facility specific disease indexes to identify patients with diagnosis of cancer. Additional methods may include reviews of pathology reports and lists of patients receiving radiation therapy or other cancer treatments to determine cases that have not been found by other methods.

Population-based registries usually depend on hospitals, clinics, radiation facilities, ambulatory surgery centers (ASC's), and pathology laboratories to identify and report cases to the central registry.

The population-based registry has a responsibility to ensure that all cases of cancer in the target area have been identified and reported to the central registry.

DATA COLLECTION FOR THE CANCER REGISTRY

Data collection methods vary between facility-based registries and population-based registries. When a case is first entered in the registry, an accession number is assigned. This number consists of the first digits of the year the patient was first seen at the facility, with the remaining digit assigned sequentially throughout the year. The first case in 2009, for example, might be 09-0001. The accession number may be assigned manually or by automated cancer database used by the organization. An **accession registry** of all cases can be kept manually or provided as a report by the database software. This listing of patients in accession number order provides a way to monitor that all case have been entered into the registry.

In a facility-based registry, data are initially obtained by reviewing and collecting them from the patient's health record. In addition to demographic information (such as name, health record number, and address), patient data in a registry include:

- Type and site of the cancer.
- Diagnostic methodologies.
- Treatment methodologies.
- Stage at the time of diagnosis.

After the initial information is collected at the patient's first encounter, information in the registry is updated periodically through he follow-up process.

Frequently, the population-based registry only collects information when the patient is diagnosed. Sometimes, however, it receives follow-up information from its reporting entities. These entities usually submit the information to the central registry electronically.

REPORTING AND FOLLOW-UP FOR CANCER REGISTRY DATA

Formal reporting of cancer registry data is done through an annual report. The annual report includes aggregate data on the number of cases in the past year by site and type of cancer. It also may include information on patients by gender, age, and ethnic group. Often a particular site or type of cancer is featured more in-depth data provided.

Other reports are provided as needed. Data from cancer registry are frequently used in the quality assessment process for a facility as well as in research. Data on survival rates by site of cancer and methods of treatment, for example, would be helpful in researching the most effective treatment for a type of cancer.

Another activity of the cancer registry is patient follow-up. On an annual basis, the registry attempts to obtain information about each patient in the registry, including whether he or she is still alive, status of the cancer, and treatment receiving during the period. Various methods are used to obtain this information. For a facility-based registry, the facility' patient health records may be checked for return hospitalizations or visits for treatment. The patient's physician also may be contacted to determine whether the patient is still living and to obtain information about the cancer. When patient status cannot be determined through these methods, an attempt may be made to contact the patient directly, using information in the registry, such as address and telephone number of the patient and other contacts. In addition, contact information from the patient's health record may be used to request information from the patient's relatives. Other methods used include reading newspaper obituaries for death and using the internet to locate patients through sites such as the social security Death Index and online telephone books.

The information obtained through follow-up is important to allow the registry to develop statistics on survival rates for particular cancers and different treatment methodologies.

Population-based registries do not always include follow-up information on the patients in their databases. Those that do, however, usually receive the information from the reporting entities such as hospitals, clinics, and other organizations providing follow-up care.

TRAUMA REGISTRIES

Trauma registries maintain databases on patients with severe traumatic injuries. A **traumatic injury** is a wound or other injury caused by an external physical force such as an automobile accident, a shooting, a stabbing, or a fall. Examples of such injuries include fractures, burns, and lacerations. Information collected by the trauma registry may be used for performance improvement and research in the area of trauma care. Trauma registries are usually facility based but may, in some cases, include data for a region or state.

CASE DEFINITION AND CASE FINDING FOR TRAUMA REGISTRIES

The case definition for trauma registry varies from registry to registry but frequently involves the inclusion of cases with diagnoses from ICD-10 of applicable range of trauma diagnoses, the trauma registrar may access the disease indexes looking for cases with codes in the applicable section of the ICD-10. In addition, the registrar may look at deaths in services with frequent trauma diagnoses such as trauma, neurosurgery, orthopedic, and plastic surgery to find additional cases.

DATA COLLECTION FOR TRAUMA REGISTRIES

After the cases have been identified, information is abstracted from the health records of the injured patients and entered into the trauma registry database. The data elements collected in the abstracting process vary from registry to registry but usually include:

- Demographic information on the patient
- Information on the injury
- Care the patient received before hospitalization (such as care at another transferring hospital or care from an emergency medical services who provided care at the scene of the accident and/or in transport from the accident site to the hospital)
- Status of the patient at the time of admission
- Patient's course in the hospital
- ICD diagnosis and procedure codes
- Abbreviated Injury Scale (AIS)
- Injury severity score (ISS)

The AIS reflects the nature of the injury and the severity (threat to life) by body system. It may be assigned manually by the registrar or generated as part of the database from data entered by the registrar. The ISS is an overall severity measurement calculated from the AIS scores for the most severe injuries of the patient. (Trauma.org 2005)

REPORTING AND FOLLOW-UP FOR TRAUMA REGISTRIES

Reporting varies among trauma registries. An annual report is often developed to show the activity of the trauma registry. Other reports may be generated as part of the performance improvement process, such as self-extubation (patients removing their own tubes) and delays in abdominal surgery or patient complication.

Trauma registries may or may not do follow-up of the patient entered in the registry. When follow-up is done emphasis is frequently on the patient's quality of life after a period of time. Unlike cancer, where physician follow-up is crucial to detect recurrence, many traumatic injuries do not require continued patient care over time. Thus follow-up is often not given the emphasis it receives in cancer registries.

BIRTH DEFECTS REGISTRIES

Birth defects registries collect information on newborns with birth defects. Often population based, these registries serve a variety of purposes. For example, they provide information on incidence of birth defects, to study causes and prevention of birth defects, to monitor trends in birth defects, to improve medical care for children with birth defects, and to target interventions for preventable birth defects such as folic acid to prevent neural tube defects.

In some cases, registries have been developed after specific events have put a spotlight on birth defects. After the Persian Gulf War, for example, some feared an increased incidence of birth defects among the children of Gulf War Veterans. The Department of Defense in the U.S. subsequently started a birth defects registry to collect data on the children of these veterans to determine whether any pattern could be detected.

CASE DEFINITION AND CASE FINDING FOR BIRTH DEFECTS REGISTRIES

Birth defects registries use a variety of criteria to determine which cases to include in the registry. Some registries limit cases to those found within the first year of life. Others include those with a major defect that occurred in the first year of life and was discovered within the first five years of life. Still other registries include only children who were liveborn or stillborn babies with discernible birth defects.

Cases may be detected in a variety of ways, including review of disease indexes, labour and delivery logs, pathology and autopsy reports, ultrasound reports, and cytogenic reports. In addition to information from hospitals and physicians, cases may be identified from rehabilitation centers and children's hospitals and from vital records such as birth, death, and fetal death certificates.

DATA COLLECTION FOR BIRTH DEFECTS REGISTRIES

A variety of information is abstracted for the birth defect registry, including:

- Demographic information
- Codes for diagnoses
- Birth weight
- Status at birth, including liveborn, stillborn, and aborted
- Autopsy
- Cytogenetics results
- Whether the infant was a single or multiple birth
- Mother's use of alcohol, tobacco, or illicit drugs
- Father's use of drugs and alcohol
- Family history of birth defects

DIABETES REGISTRIES

Diabetes registries collect data about patients with diabetes for the purpose of assistance in managing care as well as for research.

Patients whose diabetes is not kept under good control frequently have numerous complications. The diabetes registry can keep up with whether the patient has been seen by a physician in an effort to prevent complications.

CASE DEFINITION AND CASE FINDING FOR DIABETES REGISTRIES

There are two types of diabetes mellitus: insulin-dependent diabetes (type 1) and non-insulin-dependent diabetes (type 2). Registries sometimes limit their cases by type of diabetes. In some instances, there may be further definition by age. Some diabetes registries, for example, only include children with diabetes.

Case finding include the review of health records of patients with diabetes. Other case-finding methods include the reviews of the following types of information:

- ICD-10 diagnostic codes
- Billing data (if any)
- Medication lists
- Physician identification
- Health plans

Although, facility-based registries for cancer and trauma are usually hospital based, facility based diabetes registries are often maintained by clinics. The clinic is the main location for diabetes care. Thus, the data about the patient to be entered into the registry are available at the clinic or the physician office in private practice rather than at the hospital-wide level. Patient health records of diabetes patients in the clinic or physician practice may be identified through ICD-10 code number for diabetes, billing data for diabetes-related services, medication lists for patients on diabetic medications, or identification of patients as the physician sees them.

Health plans also are interested in optimal care for their enrollees because diabetes can have serious complications when not managed correctly. They may provide information to the clinic on enrollees in the health plans who are diabetics.

DATA COLLECTION FOR DIABETES REGISTRIES

In addition to demographic information about the cases, other data collected may include laboratory values such as HBA1c. This test is used to determine the patient's blood glucose for a period of approximately 60 days prior to the time of the test. Moreover, facility registries may track patient visits to follow-up with patients who have not been seen in the past years.

REPORTING AND FOLLOW-UP FOR DIABETES REGISTRIES

A variety of reports may be developed from the diabetes registry. For facility-based registries, one report may keep up with laboratory monitoring of the patient's diabetes to allow intensive intervention with patients whose diabetes is not well controlled. Another report might be of patients who have not been tested within a year or who have not had a primary care provider visit within a year.

Population-based diabetes registries might provide reporting on the incidence of diabetes for the geographic area covered by the registry. Registry data also may be used to investigate risk factors for diabetes.

Follow-up is aimed primarily at ensuring that the diabetic is seen by the physician at appropriate intervals to prevent complications.

IMPLANT REGISTRIES

An implant is a material or substance inserted in the body, such as breast implant, heart valves, and pacemakers. Implant registries have been developed for the purpose of tracking the performance of implants, including complications, deaths, and defects resulting from implants as well as longevity.

CASE DEFINITION AND CASE FINDING FOR IMPLANT REGISTRIES

Implant registries sometimes include all types of implants but often are restricted to a specific type of implant such as cochlear, saline breast, or temporomandibular joint.

DATA COLLECTION FOR IMPLANT REGISTRIES

Demographic data on patients receiving implants are included in the registry. The Food and Drug Regulatory Agency must require the following as reportable events involving medical devices:

- User facility repot number
- Name and address of the device manufacturer
- Device brand name and common name
- Product model, catalog, and serial and lot number
- Brief description of the event reported to the manufacturer and/or the regulatory agency
- Where the report was submitted

These data items also should be included in the implant registry to facilitate reporting.

REPORTING AND FOLLOW-UP FOR IMPLANT REGISTRIES

Data from the implant registry may be used to report to the regulatory agency and the manufacturer when devices or their use cause death or serious illness or injury. Follow-up is necessary to track the performance of the implant. When patients are tracked, they can be easily notified of product failure, recalls, or upgrades.

TRANSPLANT REGISTRIES

Transplant registries may have varied purposes. Some organ transplant registries maintain databases of patients who need organs. When an organ becomes available, a fair way then may be used to allocate the organ to the patient with the highest priority. In other cases, the purpose of the registry is to provide a database of potential donors for transplants using live donors, such as bone marrow transplant. Post transplant information is also kept on organ recipients and donors.

Because transplant registries are used to try to match donor's organs with recipient, they are often national or even international in scope.

Data collected in the transplant registry may also be used for research, policy analysis, and quality control projects.

CASE DEFINITION AND CASE FINDING FOR TRANSPLANT REGISTRIES

Physicians identify patients needing transplants. Information about the patient is provided to the registry. When an organ becomes available, information about it is matched with potential donors. For donor registries, donors are solicited through community information efforts similar to those carried out by blood banks to encourage blood donations.

DATA COLLECTION FOR TRANSPLANT

The type of information collected varies according to the type of registry. Pre transplant data about the recipient include:

- Demographic data
- Patient's diagnosis
- Patient's status codes regarding medical urgency
- Patient's functional status
- Whether the patient is on life support
- Previous transplantation
- Histocompatibility

Information on donors varies according to whether the donor is living. For organs harvested from patients who have died, information is collected on:

- Causes and circumstances of the death
- Organ procurement and consent process
- Medications the donor was taking
- Other donor history

For a living donor, information includes:

- Relationship of the donor to the recipient (if any)
- Clinical information
- Information on organ recovery
- Histocompatibility

REPORTING AND FOLLOW-UP FOR TRANSPLANT REGISTRIES

Reporting includes information on donors and recipients as well as survival rates, length of time on the waiting list for an organ, and death rates. Follow-up information is collected for recipients as well as living donors. For living donors, the information collected might include complications of the procedure and length of stay

(LOS) in the hospital. Follow-up on recipients includes information on status at the time of follow-up (for example, living, dead, lost to follow-up), functional status, graft status, and treatment, such as immunosuppressive drugs. Follow-up is carried out at intervals throughout the first year after the transplant and then annually thereafter.

IMMUNIZATION REGISTRIES

In the first year of life, children are supposed to receive a large number of immunizations. Immunization registries usually have purpose of increasing the number of infant and children who receive proper immunization at the proper interval. To accomplish this goal, they collect information within a particular geographic area on children and their immunization status. They also help by maintaining a central source of information for a particular child's immunization history, even when the child has received immunizations from a variety of providers. This central location for immunization data also relieves parents of the responsibility of maintaining immunization records for their own children.

CASE DEFINITION AND CASE FINDING FOR IMMUNIZATION REGISTRIES

All children in the population area served by the registry should be included in the registry. Some registries limit their inclusion of patients to those seen at public clinics, excluding those seen exclusively by private practitioners. Although, children are usually targeted in immunization registries, some registries do include information on adults for influenza, pneumonia, and other vaccines.

Children are often entered in the registry at the birth. Registry personnel may review birth and death certificates and adoption records to determine what children to include and what children to exclude because they died after birth. In some cases, children are entered electronically through a connection with an electronic birth record system. Accuracy and completeness of the data in the registry are dependent on the thoroughness of the submitters in reporting immunizations.

DATA COLLECTION FOR IMMUNIZATION REGISTRIES

The data elements to include in immunization registries include the following:

- Patient's name (first, middle, and surname)
- Patient's birth date
- Patient's sex
- Patient's birth state/country
- Mother's name (first, middle, surname, and maiden)
- Vaccine type
- Vaccine manufacturer
- Vaccination date
- Vaccine lot number

Other items may be included, as needed, by the individual registry.

REPORTING AND FOLLOW-UP FOR IMMUNIZATION REGISTRIES

Because the purpose of the immunization registry is to improve the number of children who receive immunizations in a timely manner, reporting should emphasize immunization rates, especially changes in rates in target areas. Immunization registries also can provide automatic reporting of children's immunization to schools to check the immunization status of their students.

Follow-up is directed toward reminding parents that it is time for immunizations as well as seeing whether the parents do not bring the child for the immunization after a reminder.

Reminders may include a letter or postcard or telephone calls. Autodialing systems may be used to call parents and deliver a pre-recorded reminder. Moreover, registries must decide how frequently to follow up with parents who do not bring their children for immunization. Maintaining up-to-date addresses and telephone numbers is an important factor in providing follow-up. Registries may allow parents to opt out of the registry if they prefer not to be reminded.

OTHER REGISTRIES

Registries may be developed for any type of disease or condition. Other types of registries that are commonly kept include HIV/AIDS and cardiac registries. In addition, registries may be developed for purely administrative purposes.

CHAPTER TEN

ETHICAL ISSUES IN DISEASE CLASSIFICATION AND CLINICAL CODING

INTRODUCTION

Although most people probably have never undertaken a formal study of ethics, everyone is exposed to ethical principles, moral perspective, and values throughout a lifetime. Individuals learn about basic moral values from families, religious leaders, teachers, the government, community organizations, and other groups that influence our experiences and perspectives.

There are certain actions which human being will condemn as morally wrong. The society at large frowns at embezzlers, rogues, adulterers, liars, etc, and they are said to be involved in immoral actions. However, actions like hospitality, respect for human life, honesty, and truthfulness are regarded as moral actions. (Onabajo, 2002 p. 3).[3]

According to Omoregbe (1993) ethics is concerned with fundamental principles of morality where some actions are labeled as good or bad; right or wrong; ethical or unethical and the various criteria for making such judgments.

Moral historians are of the opinion that although what is held to be right or good may vary from society to society, but the concepts of right and good are generally universal. Moral concepts changes as social life changes.

---

[3] Onabajo, O.O. (2002), Media law and Ethics.

MORAL VALUES AND ETHICAL COMPETENCIES

Ethics is the systematic study of the norms of human behavior. It is the discipline which studies the morality of human conduct and the principles of moral behavior. It is the normative science of human conduct which has the purpose of guiding human conduct along the line of moral laws. The study of ethics is to help elevate one's moral standard.

HIM professional should not make ethical decisions on behalf of others based solely on personal moral values or perspectives because not everyone share the same moral perspectives or values. Professional responsibilities often require an individual to move beyond personal values. For example, an individual might demonstrate behaviors that are based on the values of honesty, providing service to others, or demonstrating loyalty. In addition to these, professional values might require promoting confidentiality, facilitating interdisciplinary collaboration, and refusing to participate or conceal unethical practices. Professional values could require more comprehensive set of values than what an individual needs to be an ***ethical agent*** in his or her personal life. For example, an HIM professional who hears information about a friend at a party has a range of options. He or she can share the information, share only part of it, change it, or not confirm or share it with anyone. However, that same individual in his /her role as an HIM professional cannot share overheard information under any circumstances.

Ethics provides a language and a framework for formally discussing ethical issues, taking into consideration the values and obligations of others. Ethical discussion offers an opportunity to resolve conflicts when competing values are at stake.

Ethical decision making requires people to explore choices beyond to explore choices beyond the perspective of simple right or wrong (moral) options. According to Glover, "Ethics refers to the formal process of intentionally and critically analyzing the basis for one's moral judgments for clarity and consistency" (Glover 2006). When making health information decisions, HIM professionals must go beyond the personal right or wrong moral perspective and evaluate the many values and perspectives of others who are engaged in the decision to be made.

Ethical discussions outside the healthcare environment can be theoretical in nature, and the analysis of a problem does not necessarily result in an action. For example, **ethicist** could discuss whether to require all citizens living in a certain community to donate 10 hours a week to people in need as part of their duty. One ethicist might argue for a decision based on the ethical principle of beneficence, which would guide actions to do good things for others. Another ethicist might argue for the same decision but based the decision on the principle of justice in which every citizen should contribute his or her fair share for the good of the whole. These discussions and decisions would not necessarily require an action but would help frame the ethical justification for a certain action.

In contrast, **bioethics** involves problems or issues regarding clinical care or the health information system that are never strictly theoretical in nature and must always result in a decision. HIM professionals cannot merely deliberate whether to release patient information assign the correct code or purchase a new software system. Rather, they must apply ethical principles and then perform an action. In short, ethics applied in the work environment cannot remain theoretical and must result in an action.

ETHICAL FOUNDATIONS IN HEALTH INFORMATION MANAGEMENT

Ethical principles and values have been important to the HIM profession since its beginning. The first ethical pledge was presented in 1934 by Grace Whiting Myers, a visionary leader who recognized the importance of protecting information in medical records. The HIM profession was launched with recognition of the importance of privacy and the requirement of authorization for the release of health information:

I pledge myself to give out no information from any clinical record
Placed in my charge, or from any other source to any person whatsoever,
Except upon order from the Chief Executive Officer of the institution   which I may be serving (Huffman, 1972, p. 135)

Today, it is the patient who authorizes the release of information and not the chief executive officer (CEO) of the healthcare organization, as was stated in the original pledge. The most important values embedded in this pledge are to protect patient privacy and confidential information and to recognize the importance of HIM professional as a moral agent in protecting patient information (Rinehart –Thompson and Harman 2006). The HIM professional has a clear ethical and professional obligation not to give any information to anyone unless its release has been authorized, regardless of employment site, including direct patient care, facilities that access to health information, or vendors.

PROTECTION OF PRIVACY, MAINTENANCE OF CONFIDENTIALITY, AND ASSURANCE OF DATA SECURITY

The term privacy, confidentiality, and security are often used interchangeably. However, there are some important distinctions, including:

- Privacy – is "the right of an individual to be let alone. It includes freedom from observation or intrusion into one's private affairs and the right to maintain control over certain personal and health information" (Harman 2006, p. 634)
- Confidentiality – carries "the responsibility for limiting disclosure of private matters. It includes the responsibility to use, disclose or release such information only with the knowledge and consent of the individual" (Harman 2006, pp. 627-628). Confidential information may be written or verbal.
- Security – includes "physical and electronic protection of the integrity, availability and confidentiality of computer-based information and the resources used to enter, store, process, and communicate it. The means to control access and protect information from accidental or intentional disclosure" (Harman 2006, p. 635)

The HIM professional's responsibilities include ensuring that patient privacy and confidential information are protected and that data security measures are used to prevent unauthorized access to information. This responsibility includes ensuring that the release policies and procedures are accurate and up-to-date, that they are followed, and that all violations are reported to the proper authorities.

ETHICAL ISSUES RELATED TO CODING

In the past, coding was done almost exclusively for clinical studies and quality assurance review process. Although codes were provided for reimbursement purposes, the healthcare facility was reimbursed on the basis of usual, customary, and reasonable costs. The codes that were assigned became the basis of retrieval for clinical studies and the reimbursement system. Overtime, healthcare facilities have continued to use the coding systems to retrieve information in health records for clinical and administrative studies, but they also have begun to use the codes for reimbursement purposes. After the codes became the basis for reimbursement, there were inherent incentives to code so that the greatest amount of reimbursement could be given. This placed the importance of accurate coding at the fore front of the ethical issues facing HIM professionals.

Ethical problems have risen in the past few years as a result of the direct linkage between coding and payment for care. Increased pressure has been on HIM professionals who are coding to transmit inaccurate information, creating problems that are legal and/or ethical in nature. Problems include pressure to code inappropriate level of service, discovering misrepresentation in physician documentation, miscoding to avoid conflicts, discovering miscoding by other staff, lacking the tools and educational background to code accurately, and being required by employers to engage in negligent coding practices (Schraffenberger and Scichilone 2006). In response to these issues, standards have been passed that specifically address coding issues.

Failure to heed the complex rules of coding for reimbursement can lead to problems with compliance and with fraud and abuse for the HIM professional. If the HIM professional fail to establish adequate monitoring systems for accurate code assignment or submit a false claim, the consequences could include penalties, such as fees and prison. The HIM professional must know the laws and, most important, have expertise to develop preventive programmes to ensure that the false claim is non-occurrence. Fraud and abuse problems include documentation that does not justify the billed procedure, acceptance of money for information, fraudulent retrospective documentation on the part of the provider to avoid suspension, and code assignment without physician documentation. An important emerging role for the HIM professional is that of compliance officer. (Rinehart – Thompson 2006)

STANDARDS OF ETHICAL CODING

In this era of payment based on diagnostic and procedural coding, the professional ethics of health information coding professionals continue to be challenged. A conscientious goal for coding and maintaining a quality database is accurate clinical and statistical data. In the HIM world, the following are established standards of ethical coding are offered to guide coding professionals in their work:

1. Coding professionals are expected to support the importance of accurate, complete, and consistent coding practices for the production of quality healthcare data.

2. Coding professionals in all healthcare settings adhere to the ICD coding conventions, official coding guidelines and adaptations approved by cooperating partners, the International Classification of Surgical/Procedural Operations, and any official coding rules established by the WHO for use with mandated standard code sets. Selection and sequencing of diagnoses and procedures must meet the definitions of required data sets for applicable healthcare settings.

3. Coding professionals should use their skills, and their knowledge of currently mandated coding and classification systems and official resources to select the appropriate diagnostic and procedural codes.

4. Coding professionals should only assign and report codes that are clearly consistently supported by physician documentation in the health record.

5. Coding professionals should consult physicians for clarification and additional documentation prior to code assignment when there is conflicting or ambiguous data in the health record.

6. Coding professionals should not change codes or the narratives of codes on the billing abstract so that meanings are misrepresented. Diagnoses or procedures should not be inappropriately included or excluded because payment or insurance policy coverage `requirements will be affected.

When individual payer policies conflicts with official coding rules and guidelines, these policies should be obtained in writing whenever possible. Reasonable efforts should be made to educate the payer on proper coding practices in order to influence a change in the payer's policy.

7. Coding professionals as members of the healthcare team should assist and educate physicians and other clinicians by advocating proper documentation practices, further specificity, and re-sequencing or inclusion of diagnoses or procedures when needed to accurately reflect the acuity, severity, and occurrence of events.

8. Coding professionals should participate in the development of institutional coding policies and should ensure that coding policies complement, not conflict with, official coding rules.

9. Coding professionals should maintain and continually enhance their skill, as they have a professional responsibility to stay abreast of changes in codes, coding guidelines and regulations.

10. Coding professionals should strive for optimal payment to which the facility is legally entitled, remembering that it is unethical and illegal to maximize payment by means that contradict regulatory guidelines.

CHAPTER ELEVEN

FUTURE CONSIDERATION: TOWARD A BETTER TOMORROW

INTRODUCTION:

The importance of coding in the attainment of the overall objectives of healthcare industry is very central and undeniably glaring. In Nigeria, we have always recognized the central position of coding to the functions of health information management in particular, but this has always been an academic exercise, without any tangible practical and professional efforts to fully professionalize and recognize coding as a distinct specialty within the field of health information management. This is partly due to the failure of the authorities to appreciate the tremendous benefit inherent in coding and its adjoining activities.

The pre-requisite that have been set for professional venturing into coding is very rich. This is enough to make coding a very prominent specialty in HIM. This standard is meant to reflect what the role of coders entails in a well structured healthcare system. Hitherto, this has remained like the proverbial 'beautiful damsel without a suitor'.

GENERAL CODING PRE-REQUISITE

Coding combines both the theoretical knowledge and empirical observations of the coders understanding and reinforces his previous and present knowledge of the ICD. In order to master certain coding principles, the earlier ones must be mastered first and this sequence cannot be reversed. The coder must learn strategies that observe the conventions of the WHO as contained in the ICD.

Coding requires a 'logico-mathematical knowledge' which includes classification, seriation, numbering, spaces, and time concepts. The coder's role in the learning experience must be active, self-discovering, and the experience must be inductive and deductive. This is because coding and decoding involves a mental operation, which cannot be taught to students and HIM practitioners, using the most current edition of the ICD.

The coder organizes facts into conceptual systems, relates code numbers to each other and generalizes from these relationships, makes inferences and references to explain unfamiliar diseases and operations.

The mental operations in coding will often include the differentiation between a disease and an operative procedure, the identification of common properties and particular code number categories, and the determination of cause and effect relationships between principal diagnosis and disease complications. Other mental operation will include the analysis of the nature of particular coding problem or situation, retrieving relevant knowledge to determine the causal links leading to prediction or hypothesis. Using logical principles or factual knowledge to determine the necessity and sufficiency of a particular diagnosis for describing a disease condition, and determine where multiple codes are necessary. Coding is an activity that requires very close supervision in the HIM department. Where it is convenient, coders should be placed under the direct supervision of qualified and experienced HIM specialist, who must possess a good knowledge of medical terminology, nomenclature, disease classification, and current developments in the use of ICD. Where a HIM specialist places someone in charge of coding, it is still important to obtain some verification and validation of the supervisor's reliability and credibility as a coder.

One of the roles of HIM specialist is to write down the definition of terms used in the hospital or healthcare facility. Operational definitions should include the list of abbreviations used in medicine should be incorporated into the coding procedure manual for the HIM department's data processing activities.

SHORTCOMINGS OF THE ABOVE PRE-REQUISITE

The reason for the non prominence of coding activities in the country's healthcare sector is evident in the above pre-requisite. Some of them are hereby identified:

a. The eagerness of policy makers to swallow every guideline of the WHO 'hook, line, and sinker' without giving due diligence to the development of a home-grown guidelines for coding, and possibly, disease classification. In other countries of the world where coding and disease classification is taken very seriously and thereby making a lot of difference, they have taken their time to develop a home-grown classification, terminologies, and nomenclatures, (i.e. ICD-9-CM and CPT of the U.S.). This is why coding and coders are in very high demand the country's healthcare industry.

b. Most of the requirements in the pre-requisite are impracticable as long as we make coding activities an exclusive preserve of non-professionals, not until the people who have sufficient academic background in HIM take active interest in the work of coding. A visit to many of the hospitals in the country reveals the fact that coding is even non-existent.

c. Another important missing link is the absence of continuing education for coders. Although, experience matters, but the ever dynamic field of coding requires that coders are brought together for workshops, seminars and conferences, where update knowledge can be taught and shared for a positive effect and entrenchment of best practices in the coding endeavors.

d. Absence of established certification and credentialing process for coders is denigrating evidence that the country health system is not ready for serious business as far as coding is concerned.

e. No rooms for peer – review. Even there cannot be any uniformity in place as long as there are no binding national guidelines. This is also making our domestic medical researches to suffer setbacks, because they are substantially based on skeletal and inaccurate data.

THE PREFERABLE FUTURE

An ideal setting for coding cannot be any other than what obtains in the developed countries, where things work better, and they have charted a clear course and direction for disease classification system of their own in the early days of ICD development. A country that has seen it all and has been a trail blazer when it comes to HIM/ICD issues is the United States of America (U.S.A). From the following sections we want to see what an ideal setting and environment for coding should look like. The ideals, issues and examples mentioned here are those of the U.S, which have been achieved through researches, proper planning, and professional doggedness to get better as time passes by. If proper things are done a country like ours can also make a difference and join the league of HIM powers, provided there are clear visions, enough of forward looking ability, and lifelong learning.

THE NECESSARY CODING SKILLS[4]
(What Employers are looking for in coding professionals)

The employment market for coders is expected to increase 36 percent or more between 2002 and 2012, according to the Bureau of Labour Statistics. Even now we are experiencing a shortage of qualified coders. Ask a coding manager to name the biggest challenges of the job, and recruitment, and retention of qualified coders will no doubt be on the list.

This article provides insights into what employers look for from candidates. It also suggests ways new coders can increase their employability although they may lack work experience. Individual employer requirement can differ, but the themes are universal.

---

[4]   Culled from Journal of AHIMA/June 2005 – 76/6

CODING: THE BROAD VIEW

In recent years the work force landscape has changed, and the coding profession has not been immune to these changes. The image of a typical coding professional has also evolved in the eyes of employers.

Several factors influenced this change, including financial pressures on the healthcare system and advances in technology. The expectation of students entering the workforce is higher than ever, notes Gail Smith, program director of HIM at the University of Cincinnati, as there is less time to train an inexperience coder.

Coders play a vital role in translating clinical information into quality data. A coder must understand clinical diseases and invasive procedures, and it can be a challenge keeping current as medical technology advances. A coder must also understand how clinical data drive the financial aspect of healthcare. Coders need to be aware of the entire revenue cycle and compliance regulations. Facilities cannot financially afford poor data.

As healthcare services expand, so do development settings for coders. Traditionally, coders were employed by hospital; now coders work in a variety of settings ranging from physician offices to home healthcare. Coding professionals must know all laws and regulations for their specific setting.

THE SKILLS EMPLOYERS SEEK

Employers seek coding applicants knowledgeable in ICD-9-CM, CPT-4, and HCPCS (these coding books are U.S specific), though the level of knowledge required varies according to the setting. Inpatient coders may not have strong skills in CPT-4, for example, but they should have a basic knowledge of the coding system. A coder employed by a consulting company or physician office should have a strong background in physician evaluation and management coding. Coders should prepare for specific work settings by becoming familiar with the setting's coding guidelines and regulations.

Potential applicants must have other skills beyond basic coding abilities. As part of the healthcare team, coders interact with a variety of healthcare professionals from the business office to clinicians. Thus it is important that they possess effective interpersonal communication skills. Written communication skills are also very vital to the coding profession. Employers must ensure that coders can form an appropriate physician query.

Coders must embrace new technology as well as be willing to learn new software and solve technical issues. Time management skills are also a necessity, especially as coders move into the realm of remote coding. Employers look for individuals who take initiative to research references and solve difficult cases.

EDUCATION: A NECESSITY

The preference for an applicant's degree of education may vary according to setting. However, education is still a top priority for employers looking for coding professionals. Employers want applicants who have had basic coding classes, at minimum. Employers may desire that the applicants complete an AHIMA certification program at a minimum.

Because of the shortage of qualified coders, some employers may hire applicants who have had training in other allied health fields and have taken formal coding classes, even if not in an approved *program. Employers prefer that candidates have credentials, but they may not require them. They may also be willing to hire a candidate in the process of earning credentials or may make pursuing credentials a condition of employment.*

Education may not involve a formal institution. Deb Boppre, VA project manager at United Audit Systems, notes that the more knowledge and experience you have coding in different settings, the more marketable you are as a coder. If you are currently coding one specific type of case – for example paediatrics – perform a self-assessment to see what other coding skills may need updating. Investigate ways to obtain or improve in the area. For example, you could take a continuing education class online or attend a local seminar.

LOCATING EMPLOYMENT OPPORTUNITIES

While there is no one way for employer to recruit, a potential applicant should take note of some effective methods employers have used. One recruitment technique is a partnership with area HIM programs. An internship for coders from various programs to work 80 hours over a period of time was an effective way to fill open coding positions. Susan Schehr, chief of HIM at the Department of Veterans Affairs in Cincinnati, OH, explains that students took actual records to code and worked with a coder. Currently, employed coders, students, and the manager met at a weekly roundtable to discuss the cases and review the student's work. While the employer did not pay the interns, Schehr says, the facility did hire two of them to fill coding vacancies.

Networking and referrals are other effective ways to find open coding positions. Become involved in HIM association in the area where you wish to obtain employment. Do not be afraid to get involved. Start by volunteering for your local association. Not only does this provide you with a way to network, but it also displays your skills as an HIM professional to potential employers by working side by side with them.

INTERVIEWING WITH POTENTIAL EMPLOYERS

The method by which employers assess an applicant's skills varies, but one common tool is a comprehensive coding exam. Employers typically ask applicants a series of behavior-based questions to assess skills after exam. Do not be surprised if the questions include defining coding terms. Collette Ferguson, assistant chief of HIM department of Veterans Affairs in Cincinnati, OH, notes she has interviewed numerous candidates who have struggled with basic definitions of coding-related terms, such as identifying three key components of an evaluation and management code.

Another factor influencing an employer's decision on hiring is references provided by the applicant. References should be professionally specific, and you should inform your references that they might be called so that they are prepared when a potential employer contacts them.

Higher employer expectations have led to changes in coding education. As employers have come to expect higher-level skills from graduates, instructors have responded by providing students with some real life cases. Students should prepare portfolios of successfully completed classroom projects. During interviews, these may be used to illustrate learned skills applied to actual cases.

Do not expect to be new to coding profession and land your perfect job right away. Be open to interviewing in various settings. You never know where the opportunity may lead. Showcase your skills during the interview process.

As the coding profession continues to change, it is important for coding professionals to pursue lifelong learning. Commitment to ongoing education will lead to endless professional opportunities.

BEYOND CODING TO CONTENT ANALYSIS[5]

The **Health Information Technology** (HIT) field can provide HIM professionals with many career opportunities at software and content companies. HIT roles require highly developed content analysis skills, including abstracting, coding, documentation, compliance, and utilization review. The availability of rewarding healthcare careers for content analysts represents one of the best kept secrets of the healthcare industry.

Working in HIT (e.g. a provider of electronic medical records or electronic decision support content) requires additional skills for HIM professionals. HIT vendors focus on creating content that covers abroad range of healthcare subjects. Content analysts work with content and data for large groups of patients or providers rather than on an individual patient basis.

This section describes the role of the content analyst, outlining some of the required skills and qualifications. A summary of the content analyst's role relating to other HIT professionals is also

---

[5] Culled from Journal of AHIMA/June 2005 – 76/6

discussed. Finally, a general description of the different kinds of HIT companies is included.

THE CONTENT ANALYST

Medical content in the healthcare industry includes many comprehensive vocabularies. HIM professionals currently use ICD, CPT, and HCPCS (i.e. in the U.S) for medical content analysis. In the HIT world, content analysts deal with a wide range of clinical vocabularies such as SNOMED-CT, LOINC, Rx Norm, and others.

At HIT companies, content not only include vocabularies but also drug information, decision support such as drug-interaction information, and monograph information for patients and physicians. HIM professionals working in HIT have an opportunity to learn terminologies and content domains outside of the standard administrative code sets.

The qualification to become a content analyst includes expert knowledge of medical terminology, anatomy, and physiology. A content analyst is expected to be knowledgeable on all code sets for classification of diseases, procedures, billing, and reimbursement. Proficiency in the use of computers and software such as word processing and spreadsheets is a necessity. Additional computer skill such as database querying and programming are not required but may be advantageous.

Responsibilities of a content analyst include mapping and modeling code sets (i.e. ICD, CPT, HCPCS, and APC) to controlled medical vocabularies such as SNOMED-CT. Other tasks performed by the content analyst include:

- Grouping related clinical terms into lists to support structured data entry. Based on billing and medical knowledge, content analysts choose the most common diagnoses and procedures to include in the lists.

- Reviewing billing revenue cycles from official sources and adding the information to software applications for customer use. This information is reviewed daily to meet the ever-changing billing and reimbursement guidelines.

- Reviewing and contacting information sources such as the centers for Medicare and Medicaid Services (an equivalent of Nigeria's NHIS), the Federal Register, the Medical Association, and the WHO.

- Writing documentation for all content releases. Technical writing skills are required.

- Using and understanding software used in data capture and communication such as Microsoft Office applications as well as other applications, some of which are homegrown.

GROWING INTO OTHER ROLES

Opportunities exist for HIM professionals to expand into project management and customer service roles as they become familiar with the role of the content analyst. Project managers plan, direct, and coordinate activities of designated projects to ensure that goals or objectives of a project are accomplished within the prescribed time frame. It is critical for project managers to maintain a list of all current content projects and their deadlines, manage tool needs and issues, ensure that customer documentation is accurate and easily understood, and formalize methods to capture content changes and any corrections that must be made. Good customer interaction skills go hand in hand with this role, as project managers must be able to communicate with customers to quickly identify and solve conflicts while maintaining high client satisfaction.

In addition to opportunities in analyst roles, HIM professionals can also become involved in training customers in the use of the tools and content offered by HIT companies. Training involves communication skills and considerable interaction with current and potential customers.

THE HIT TEAM

HIT vendors that specialize in the delivery of content and terminologies require a wide range skill sets. In this age of evolving standards and a wide audience for terminologies, a content team within software company needs traditional content domain expertise as well as software engineers. Medical and nursing informaticist who have both clinical and computing experience and training are often required in order to understand the use and implementation of terminologies and other contents within actual systems. Informaticist also play a large role in defining the content needs of users such as electronic health record companies and governments and in developing new kinds of content such as mapping among terminologies. Content analysts work closely with informaticist in HIT companies.

The actual delivery of content to users often requires a different set of skill sets; engineering and development. Software tools are required in order to build and maintain content. For example, when developing mappings among terminologies, a software tool should allow users to assign work lists, perform the actual mappings, tracking the work history, and perform quality assurance testing. Software developers are required to create and maintain the tools as well as the various projects within them. Content analysts work with software developers to design the requirements for the tools and also to report on bugs and maintenance issues.

Content delivery to users requires engineering in order to package the correct content that has been tested into the appropriate computerized formats. Expertise in software development and engineering skills such as database creation, maintenance, querying, and delivering are required. Once the content is packaged into the software, the content analyst's skills are again required to certify that the content is correct. Often, a content analyst with domain expertise is the only individual who can determine whether the content is correctly loaded into the software.

THE HIT INDUSTRY

The HIM professional's expertise in the use of coding terminologies is a natural fit in HIT companies that focus on billing applications. For example, companies that create billing and practice management software build applications to assist clinicians, hospitals, and offices in billing electronically. But with the expanded role of clinical terminologies such as SNOMED-CT, opportunities exist in several other kinds of HIT companies. For example, electronic health records vendors create applications that present terminologies, including billing codes, to end users such as physicians and nurses. The content analyst can help these vendors transition from presenting billing codes to clinical terminology codes to end users.

Many HIT companies specialize in content such as drug information, patient education leaflets, and disease information. These content companies often require domain expertise from physicians, nurses, and pharmacist, but HIM professionals have skills to help map and index this content to billing standards. In addition, HIM professionals can leverage expertise in anatomy and other specialties to grow into a content author role for these types of companies.

HIM professionals can take parts in creating cutting-edge technology, making quantum leaps in healthcare delivery. The collection and use of medical data and knowledge is critical to the practice of good and safe healthcare. Content analysts with HIM backgrounds can make the difference now and in the future by working in the field of HIT.

ABBREVIATIONS

ACS – American College of Surgeons

AHIMA – American Health Information Management Association

AI – Artificial Intelligence

ALOS – Average Length of Stay

AMA – American Medical Association

ANA – American Nurses Association

ANSI – American National Standard Institute

APC – Ambulatory Payment Classification

APS – Attending Physician Statement

ASP – Application Service Provider

CDR – Clinical Data Repository

CDS – Clinical Data Support

CDSS – Clinical Decision Support System

CDT – Current Dental Terminology

CDW – Clinical Data Warehouse

CEO – Chief Executive Officer

CMS – Centers for Medicare and Medicaid Services

CPT – Current Procedural Terminology

DICOM – Digital Imaging and Communication in Medicine

DRG – Diagnosis – Related Groups

Dx – Diagnosis

EHR – Electronic Health Record

EMR – Electronic Medical Record

EOC – Episode of Care

ES – Expert System

FDA – Food and Drug Administration

HCPCS – Healthcare Common Procedure Coding System

HIE – Health Information Exchange

HIM – Health Information Management

HIS – Hospital Information System

HIT – Health or Healthcare Information Technology

HMIS – Health Management Information System

HMO – Health Maintenance Organization

ICD-9 – International Statistical Classification of Diseases, Injury and Causes of Death, Ninth Revision

ICD-10 – International Statistical Classification of Disease and Related Health Problems, Tenth Revision

ICD-9-CM – International Classification of Diseases, Ninth Revision, Clinical Modification

ICD-10-CM – International Classification of Diseases, Tenth Revision, Clinical Modification

ICD-O – International Classification of Diseases for Oncology

ICF – International Classification on Functioning, Disability, and Health

ICNP – International Classification for Nursing Practice

ICPC-2 – International Classification of Primary Care

IMIA – International Medical Informatics Association

IOM – Institute of Medicine

IS - Information System

ISC – International Statistical Congress

ISI – International Statistical Institute

ISO – United Nations International Standards Organization

LOINC – Logical Observation Identifier Names and Codes

LOS – Length of Stay

MCO – Managed Care Organization

MEDLINE – Medical Literature, Analysis, and Retrieval System Online

NAHIT – National Alliance for Health Information Technology

NANDA – North American Nursing Diagnosis Association

NCHS – National Center for Health Statistics

NCRA – National Cancer Registrars Association

NCVHS – National Committee on Vital and Health statistics

NDC – National Drug Codes

NIC – Nursing Intervention Classification

NISDEC – Nursing Information and Data Set Evaluation Center

NLM – National Library of Medicine

NLP – National Language Processing

NMDS – Nursing Minimum Data Set

NMMDS – Nursing Management Minimum Data Set

NOC – Nursing Outcomes Classification

PDA – Personal Digital Assistant

SNOMED – Systematized Nomenclature of Medicine

SNOMED CT – Systematized Nomenclature of Medicine Clinical Terms

SNOMED RT – Systematized Nomenclature of Medicine Reference Terminology

SSN – Social Security Number

UHDDS – Uniform Hospital Discharge Data Set

UMDNS – Universal Medical Device Nomenclature

UMLS – Unified Medical Language System

WHO – World Health Organization

GLOSSARY

Abbreviated Injury Scale (AIS): A set of numbers used in a trauma registry to indicate the nature and severity of injuries by body system

ABC Codes: A terminology created by Alternative Link that describes alternative medicine, nursing, and other integrative healthcare interventions

Accession Number: A number assigned to each case as it is entered in a cancer registry

Accession registry: A list of cases in a cancer registry in the order in which they were entered

Acute care prospective payment system (PPS): The reimbursement system for inpatient hospital services provided to Medicare and Medicaid beneficiaries that is based on the use of diagnosis-related groups as a classification tool

Aggregated data: data extracted from individual health records and combined to form de-identified information about groups of patients that can be compared and analyzed

Ambulatory care: Preventative or corrective healthcare services provided on a non resident basis in a provider's office, clinic setting, or hospital outpatient setting

Ambulatory Payment Classification (APC) System: The Prospective payment system used in the United States since 2000 for reimbursement of hospitals for outpatient services provided to Medicare and Medicaid beneficiaries

American Health Information Management Association (AHIMA): The professional membership organization for managers of health record services and healthcare information system as well as coding services; provides accreditation, certification, and educational services

Autocoding: The process of extracting and translating dictated and then transcribed free-text data (or dictated and then computer-generated discrete data) into ICD-9-CM and CPT evaluation and management codes for billing and coding purposes

Average Length of Stay (ALOS): The mean length of stay for hospital inpatients discharged during a given period of time

Benchmarking: An analysis process based on comparison

Bill of Mortality: Documents used in London during the seventeenth century to identify the most common causes of death

Bioethics: A field of study that applies ethical principles to decisions that affect the lives of humans, such as whether to approve or deny access to health information

Case definition: A method of determining criteria for cases that should be included in a registry

Case finding: A method of identifying patients who have been seen and/or treated in a healthcare facility for the particular disease or condition of interest to the registry

Classification: A clinical vocabulary, terminology, or nomenclature that lists words or phrases with their meanings, provides for the proper use of clinical words as names or symbols, and facilitates mapping standardized terms to broader classifications for administrative, regulatory, oversight, and fiscal requirements

Classification System: 1. A system for grouping similar diseases and procedures and organizing related information for easy retrieval   2. A system for assigning numeric or alphanumeric code numbers to represent specific diseases and/or procedures

Clinical Care Classification (CCC): Two interrelated taxonomies, the CCC of Nursing Diagnoses and Outcomes and the CCC of Nursing Interventions and Actions, that provide a standardized framework for documenting patient care in hospitals, home health agencies, ambulatory care clinics, and other healthcare settings

Clinical data: Data captured during the process of diagnosis and treatment

Clinical data repository (CDR): A central database that focuses on clinical information

Clinical Decision Support System (CDSS): A special subcategory of clinical information systems that is designed to help healthcare providers make knowledge-based clinical decisions

Coded data: Data that are translated into a standard nomenclature of classification so that they may be aggregated, analyzed and compared

Coding: The process of assigning numeric/alphanumeric representations to clinical documentation

Co-morbidity: A medical condition that coexists with the primary cause for hospitalization and affects the patient's treatment and length of stay

Complication: A medical condition that arises during an inpatient hospitalization (for example, a postoperative wound infection)

Confidentiality: A legal and ethical concept that establishes the healthcare provider's responsibility for protecting health records and other personal and private information from unauthorized use or disclosure

Data set: A list of recommended data elements with uniform definitions that are relevant for a particular use

Discharge summary: A summary of the resident's stay at the long-term care facility that is used along with the post-discharge plan of care to provide continuity of care for the resident upon discharge from the facility

Disease index: A list of diseases and conditions of patients sequenced according to code numbers of the classification system in use

Disease registry: A centralized collection of data used to improve the quality of care and measure the effectiveness of a particular aspect of healthcare delivery

Disposition: For outpatients, the healthcare practitioner's description of the patient's status at discharge (no follow-up planned; follow-up planned or scheduled; referred elsewhere; expired); for inpatients, a core health data element that identifies the circumstances under which the patient left the hospital (discharged alive; discharged to home or self care; discharged and transferred to another short-term general hospital for inpatient care; discharged and transferred to a skilled nursing facility; discharged and transferred to an intermediate care facility;

discharged and transferred to another type of institution for inpatient care or referred for outpatient services to another institution; discharged and transferred to home under care of organized home health services organization; discharged and transferred to home under care of a home intravenous therapy provider; left against medical advice or discontinued care; expired; status not stated)

Effectiveness: The degree to which stated outcomes are attained

Efficiency: The degree to which a minimum of resources is used to obtain outcomes

e-HIM: The application of technology to managing health information

Ethical agent: An individual who promotes and supports ethical behaviour

Ethicist: An individual trained in the application of ethical theories and principles to problems that cannot be easily solved because of conflicting values, perspectives and options for action

Ethics: A field of study that deals with moral principles, theories, and values; in healthcare, a formal decision-making process for dealing with the competing perspectives and obligations of the people who have an interest in a common problem

Ethics training: The act of teaching others about moral principles, theories, and values

Health Information Management (HIM): An allied health profession that is responsible for ensuring the availability, accuracy, and protection of the clinical information that is needed to deliver healthcare services and to make appropriate healthcare-related decisions

ICD-9-CM (International Classification of Diseases, Ninth Revision, Clinical Modification): A classification system used in the United States to report morbidity and mortality information

Index: An organized (usually alphabetical) list of specific data that serves to guide, indicate or otherwise facilitate reference to the data

International Classification for Nursing Practice (ICNP®): Unified nursing language system into which existing terminologies can be cross-mapped

International Classification of Diseases for Oncology (ICD-O): A classification system used for reporting incidences of malignant disease

International Classification of Primary Care (ICPC-2): Classification used for coding the reasons for encounter, diagnoses, and interventions in an episode-of-care structure

International Classification on Functioning, Disability, and Health (ICF): Classification of health and health related domains that describe body functions and structures, activities, and participation

Nomenclature: A recognized system of terms used in a science or art that follows pre-established naming conventions; a disease nomenclature is a listing of the proper name for each disease entity with its specific code number

Nosology: The branch of medical science that deals with classification systems

Personal digital assistant (PDA): A hand-held micro-computer, without a hard drive, that is capable of running applications such as e –mail and providing access to data and information, such as notes, phone lists, schedules, and laboratory results, primarily through a pen device

Principal diagnosis: The disease or condition that was present on admission, was the principal reason for admission, and received treatment or evaluation during the hospital stay or visit

Principal procedure: The procedure performed for the definitive treatment of a condition (as opposed to a procedure performed for diagnostic or exploratory purposes) or for care of a complication

Privacy: The quality or a state of being hidden from, or undisturbed by, the observation or activities of other persons or freedom from unauthorized intrusion; in healthcare-related contexts, the right of a patient to control disclosure of personal information

Public health: An area of healthcare that deals with the health of populations in geopolitical areas, such as states and counties

Quality: The degree or grade of excellence of goods or services, including, in healthcare, meeting expectations for outcomes of care

Quality assurance (QA): A set of activities designed to measure the quality of a service, product, or process with remedial action, as needed, to maintain a desired standard

Speech recognition technology: Technology that translate speech to text

Systematized Nomenclature of Medicine (SNOMED): A comprehensive clinical vocabulary developed by the College of American Pathologists that is the most promising set of clinical terms available for a controlled vocabulary for healthcare

Systematized Nomenclature of Medicine Clinical Terminology (SNOMED CT): A comprehensive, controlled clinical vocabulary developed by the College of American Pathologists

Systematized Nomenclature of Medicine Reference Terminology (SNOMED RT): A concept-based terminology consisting of more than 110,000 concepts with linkages to more than 180,000 terms with unique computer-readable codes

Taxonomy: The principles of a classification system, such as data classification, and the study of the general principles of scientific classification

Terminology: A set of terms representing the system of concepts of a particular subject field; a clinical terminology provides the proper use of clinical words as names or symbols

Vital statistics: Data related to births, deaths, marriages, and fetal death

Vocabulary: A list or collection of clinical words or phrases and their meaning

Vocabulary standard: A common definition for medical terms to encourage consistent descriptions of an individual's condition in the health record

BIBLIOGRAPHY

BOOKS

Bowman, E. 2001. Coding, Classification, and Reimbursement Systems. In Health Information: Management of a Strategic Resource, 2nd ed, edited by M. Abdelhak et al. Philadelphia: W.B. Saunders Company.

Dochterman, J. C., and G.M. Bulechek (Eds.). 2004. Nursing Interventions Classification (NIC), fourth edition. St Louis, MO: Mosby.

Eichenwald-Maki, S., and K.M. LaTour (Eds) 2006. Health Information Management: Concepts, Principles, and Practice, 2nd ed: Chicago,IL: AHIMA

Fenton, S. H., and M. Greene. 2006. Clinical Classifications and Terminologies. Chapter 13 in Health Information Management: Concepts, Principles, and Practice, 2nd ed, edited by Eichenwald-Maki, S., and K.M. LaTour: Chicago,IL: AHIMA

Glover, J. J., 2006. Ethical decision-making guidelines and tools. Chapter 2 in Ethical Challenges in the Management of Health Information: Process and Strategies for Decision-making, 2nd edition, edited by L. B. Harman: Sudbury, M. A: Jones and Bartlett.

Hammond, W. E., and J. J. Cimino. 2000. Standards in Medical Informatics. In Medical Informatics: Computer Applications in Healthcare and Biomedicine, 2nd ed, edited by E.H. Shortliffe and L.E. Perreault, pp. 212-56. New York: Springer.

Harman, L. B., ed. 2006. Ethical Challenges in Management of Health Information: Process and Strategies for Decision-making. 2nd ed. Sudbury MA: Jones and Bartlett.

Huffman, E. K., 1972. Manual for Medical Records Librarians, 6th ed. Chicago: Physician's Record Company.

Neuberger. B. J. 2006. Public Health. Chapter 8 in Ethical Challenges in the Management of Health Information: Process and Strategies for Decision-making, 2nd edition, edited by L. B. Harman: Sudbury, M. A: Jones and Bartlett.

Rinehart – Thompson, L. A. 2006. Compliance, fraud, and Abuse. Chapter 4 in Ethical Challenges in the Management of Health Information: Process and Strategies for Decision-making, 2nd edition, edited by L. B. Harman: Sudbury, M. A: Jones and Bartlett.

Rinehart – Thompson, L. A., and L. B. Harman. 2006. Privacy and Confidentiality. Chapter 3 in Ethical Challenges in the Management of Health Information: Process and Strategies for Decision-making, 2nd edition, edited by L. B. Harman: Sudbury, M. A: Jones and Bartlett.

Schraffenberger, L. A., and R.A. Scichilone. 2006. Clinical code selection and use. Chapter 5 in Ethical Challenges in the Management of Health Information: Process and Strategies for Decision-making, 2nd edition, edited by L. B. Harman: Sudbury, M. A: Jones and Bartlett.

International Statistical Classification of Diseases and Related Health Problems, Tenth Revision.1993: Vol. 1-3: Geneva: WHO

JOURNALS

Barnett, O. G., et al. 1993. The computer-based clinical records: where do we stand? Annals of Internal Medicine 119(10): 1046-48.

Beinborn, J. 1999. Automated Coding: The next step. Journal of American Health Information Management Association 70(7)

Bronnert, J. 2005. The Necessary Coding Skills: what employers are looking for in coding professionals. Journal of American Health Information Management Association 76/6: 60-61

Chute, C. G. 2000. Clinical Classification and Terminology: Some history and current observations. Journal of American Health Information Management Association 70 (3): 298-303.

Cimino, J. J. 1998. Desiderata for controlled medical vocabularies in the twenty-first century. Methods of Information in Medicine 37: 394-403.

Forrey, A.W., C. J. McDonald, G. DeMoor, et al. 1996. Logical Observation Identifier Names and Codes (LOINC) Database: A public use set of codes and names for electronic reporting of clinical laboratory results. Clinical Chemistry 42: 1: 81-90.

Garthe, E. 1997. Overview of trauma registries in the United States. Journal of American Health Information Management Association 68 (7): 26, 28, 30-31Henry, S. B., et al. 1998. Nursing data, classification systems, and quality indicators: what every HIM professional needs to know. Journal of American Health Information Management Association 69 (5): 48-54

Kudla, K.M., and M. Blakemore. 2001. SNOMED takes the next step. Journal of American Health Information Association 72 (7): 62-68.

Richmond, R. 2005. Beyond Coding to Content Analysis. Journal of American Health Information Management Association 76/6: 66-67

WEBSITE (INTERNET)

American College of Surgeons, 2005. Trauma Programs. Available online from facs.org/cancer/

American National Standard Institute, n.d.ansi.org

Center for Disease Controls. 2005. National Immunization Program. Available online from cdc.gov/nip/ registry/min-funct-stds 2001.html

Food and Drug Administration. n.d fda.gov

International Classification for Nursing Practice (ICNP). 2004. Available online from icn.ch/icnp.htm

Jamoulle, M. 1998. ICPC-2: The International Classification of Primary Care: an introduction. Available online from ulb.ac.be/esp/wicc/icpc2.html.

Martin, K.S. 2005. Omaha System. Available online from omahasystem.org

National Alliance for Health Information Technology (NAHIT) 2005. Alliance Standards Directory. Available online from hitsdir.org

National Cancer Registrars Association. 2005. ncra-usa.org

National Center for Health Statistics. ICD-9. Available online from cdc.gov.org/nchs

National Center for Health Statistics, ICD-10-CM. Availlable online from cdc.gov/nchs

Ringer, D., and W. Cain. 2004. COC revises cancer standards, creates FORDS. Advance for Health Information Professionals. Available online from health-information.advanceweb.com/common/Editorial Search/A viewer. Aspx/cc = 28947

Trauma.org. 2005

About the Book

A Modern Approach to Disease Classification and Clinical Coding is a paradigm Shift in the field of Health Information Management (HIM) in general. For more than three decades that serious efforts at providing academic instructions in Health Record Management started in Nigeria, the content of the curriculum for Disease Classification had not seen so much change, the same content taught in the 1970's and 1980's is still what students are being taught presently. The only new thing the instructors pass to the students is the new revisions of the ICD, which come out once in ten years.

The good thing about this book is that it contains virtually all the up-to-date information and best practices in Disease Classification and Clinical Coding, ranging from recently developed classification systems, updates in ICD-10, and technological and ethical applications in coding, with a glimpse in to the future of this specialty of HIM.

About the Author

Folasayo Ayegbayo is currently the Head, School of Continuing Education, Lagos State College of Health Technology, Yaba. He was formerly the Head of Department of Health Information Management of the College. With almost a two decades of classroom experience, both at home and other Neighbouring West African Countries.

An award-winning alumnus of University College Hospital, Ibadan; Lagos University Teaching Hospital, Idi-Araba; Ambrose Alli University, Ekpoma; Houdegbe North American University, Benin Republic; Ecole Supérieure de Commerce et d'Administration de Entreprises (ESCAE – Porto Novo, Benin); CIML Business School, Delaware USA. A Fellow of the Chartered Institute of Management and Leadership. He has been teaching Disease Classification and Clinical Coding since the beginning of his teaching career.

He is married with children.